# Conservation through Sustainable Use

The human use of nature is a polarizing topic in India and across the globe, often perceived as contradictory to traditional exclusionary conservation. However, India's natural landscapes serve as important sources of biological resources for many communities. This collection of case studies on sustainable use practices throughout India aims to identify the policies, management strategies, and knowledge contexts that contribute to resource use without damaging biological diversity.

Through a diverse array of personal accounts, stories and photographs from the field, and ongoing research studies across biogeographic zones, readers will connect with academics, practitioners, managers, and policy analysts who challenge us to rethink the conservation paradigm. These chapters provide a reflection on the history of conservation and sustainable use in India and illuminate a path towards a local and global future in which biodiversity and human well-being go hand in hand.

The wide variety of authors in this book reflects the broad audience this book will be of interest to, from students studying environmental conservation and sustainability to researchers, practitioners, and policymakers who work in the field and seek to learn about successful sustainable use systems and resulting lessons that have widespread application. This book will appeal to readers interested in the areas of environment sciences, biodiversity management, sustainable development, developmental studies, forestry, wildlife and protected area management, public policy, environmental policy, and governance.

**Anita Varghese** is Director (Biodiversity) at Keystone Foundation. She holds a bachelor's degree in Zoology (Bombay University), master's degree in Ecology (Pondicherry University), and a doctorate in Botany (University of Hawaii). Her interests are in plant conservation specifically on sustainable use, non-timber forest products, long-term population dynamics of harvested species, invasive plants, cycads, and forest trees.

**Meera Anna Oommen** is a trustee of Dakshin Foundation, Bangalore, and the Madras Crocodile Bank. She works on issues related to ecology, conservation science, and environmental history. Her recent work focuses on incorporating insights from multiple disciplines to understand the dynamics of human–wildlife conflict, aspects related to human–animal relationships, and the history of hunting in India.

**Mridula Mary Paul** is a postgraduate researcher at the Department of Geography and Environmental Sciences, Northumbria University, Newcastle-upon-Tyne, UK. She has previously been Senior Policy Analyst with the Ashoka Trust for Research in Ecology and the Environment (ATREE), Bengaluru, and practised environmental law before the Madras High Court.

**Snehlata Nath** is Founder Director of Keystone Foundation and has worked on conservation-livelihoods-enterprise with indigenous people in the Nilgiri Biosphere Reserve. She co-founded the NTFP Exchange Program network across Asia and has coordinated the India chapter for over 20 years.

# Conservation through Sustainable Use

## Lessons from India

Edited by Anita Varghese, Meera Anna Oommen, Mridula Mary Paul, and Snehlata Nath

Routledge
Taylor & Francis Group
LONDON AND NEW YORK

First published 2023
by Routledge
4 Park Square, Milton Park, Abingdon, Oxon OX14 4RN

and by Routledge
605 Third Avenue, New York, NY 10158

*Routledge is an imprint of the Taylor & Francis Group, an informa business*



*British Library Cataloguing-in-Publication Data*
A catalogue record for this book is available from the British Library

ISBN: 978-1-032-29083-6 (hbk)
ISBN: 978-1-032-38102-2 (pbk)
ISBN: 978-1-003-34349-3 (ebk)

DOI: 10.4324/9781003343493

Typeset in Sabon
by Apex CoVantage, LLC

# Contents

# Figures

# Contributors

**Anita Varghese** is Director (Biodiversity) at Keystone Foundation. She holds a bachelor's degree in Zoology (Bombay University), master's degree in Ecology (Pondicherry University), and a Doctorate in Botany (University of Hawaii). Her interests are in plant conservation specifically on sustainable use, non-timber forest products, long-term population dynamics of harvested species, invasive plants, cycads, and forest trees. Her work looks at the factors that mediate the relationship between people and nature, specifically how the goals of conservation and development can be harmonized. She is Chair of the Western Ghats Plant Specialist Group of the SSC IUCN. Additionally she is a member of the Sustainable Use and Livelihoods Specialist Group steering committee (CEESP/SSC–IUCN) and Cycad specialist group (SSC/IUCN).

**Mridula Mary Paul** is a postgraduate researcher at the Department of Geography and Environmental Sciences, Northumbria University, Newcastle-upon-Tyne, UK. She has previously been Senior Policy Analyst with the Ashoka Trust for Research in Ecology and the Environment (ATREE), Bengaluru, and practised environmental law before the Madras High Court. Mridula has a degree in Development Studies from the University of Oxford. She currently works on issues of biodiversity and health, specifically focusing on the political ecology of zoonoses and One Health. She is the editor of *Courting the Environment*, a newsletter that aims to convey environmental and ecological research to lawyers.

**Meera Anna Oommen** is a trustee of Dakshin Foundation, Bangalore, and the Madras Crocodile Bank. She works on issues related to ecology, conservation science, and environmental history. Her recent work focuses on incorporating insights from multiple disciplines to understand the dynamics of human–wildlife conflict, aspects related to human–animal relationships, and the history of hunting in India. Her interests also lie in integrating emerging research in the fields of frugal heuristics, bounded rationality, and artificial intelligence in the context of conservation. Her working experience has largely been in mountain (Trans-Himalaya,

Western Himalaya, and the Western Ghats) and island systems (Andaman and Nicobar Islands).

**Snehlata Nath** is Founder Director of Keystone Foundation and has worked on conservation-livelihoods-enterprise with indigenous people in the Nilgiri Biosphere Reserve. She co-founded the NTFP Exchange Program network across Asia and has coordinated the India chapter for over 20 years. This has a membership of over 80 NGOs working with indigenous people and forests. She has established community collectives on NTFP-based livelihoods and advocates for the rights over resources of indigenous people. Traditional knowledge, sustainable use, and community-led conservation with a gender lens are themes of her interest.

**Naveen Namboothri** is a founder trustee of Dakshin Foundation and heads the Sustainable Fisheries programme at Dakshin. Trained as a marine biologist, he has worked in diverse coastal and marine systems across India particularly focusing on its island systems. Naveen's research interests range widely from natural history of marine organisms to understanding the interrelationships between humans and marine ecosystems to developing participatory and equitable resource management regimes in India. He has a PhD in Marine Biology.

**Ishaan Khot** is a marine biologist by training and his work focuses on the human dimensions of the oceans. Following a master's degree in Marine Biology from Pondicherry University, Ishaan started working with Dakshin in 2015 on community-based fisheries monitoring and management initiatives in the Lakshadweep Islands. In his current capacity as the Programme Officer of the Sustainable Fisheries programme, he coordinates and oversees Dakshin's interdisciplinary research and intervention projects on sustainable fisheries management in field sites across various coastal geographies in India.

**Abel Job Abraham** is formally trained in subjects such as Criminology, Child Rights Law, and Development Studies. His research interests lie in exploring various development issues from a rights-based perspective using an interdisciplinary approach. Abel currently studies the usage of commons and the legally pluralistic systems of governance on Minicoy Island as a part of Dakshin's larger work towards creating a rights-based, participatory model of fisheries co-management in the Lakshadweep Islands.

**Mr. Jagannatha Rao R.** holds a master's degree in Agricultural Sciences. He is working as a faculty at The University of Trans-Disciplinary Health Sciences and Technology, Bengaluru, India. He has worked on ecosystem services offered by natural landscapes, designing and conducting value chain assessment studies for NTFPs/medicinal plants, sustainable wild harvesting, cultivation, value addition and marketing, training and capacity-building programmes, and is involved in standardizing field collection practices for International Standards for Sustainable Collection of Medicinal and Aromatic plants (ISSC-MAP). He has a number of publications in journals and books. He has edited a book titled 'Adaptive management of NTFPs/medicinal plants: Strategies, Implications and Policy'.

**Ms. Deepa G.B.** is a postgraduate in Environmental Sciences, working as a Faculty at The University of Trans-Disciplinary Health Sciences and Technology, Bengaluru, India. She has worked on field implementation of good collection practices, which includes wild collection, cultivation, semi processing, value addition and marketing, and regularly conducts community training and skill development programmes for different stakeholders on sustainable utilization of medicinal plants. She has developed a concept note on Principles and Practices of Sustainable Use and Sustainable Harvesting based on the vast field experience of implementation of sustainable harvesting practices in India and has a number of publications.

**Ambika Aiyadurai** is Assistant Professor (Anthropology) in the Indian Institute of Technology – Gandhinagar. She is an anthropologist of wildlife conservation with a focus on human–animal relations and community-based conservation. She completed her PhD thesis in Anthropology from the National University of Singapore in 2016. She is trained in natural and social sciences with masters' degrees in Wildlife Sciences from Wildlife Institute of India (Dehradun) and Anthropology, Environment and Development from University College London (UK) funded by Ford Foundation's International Fellowship Program. Her monograph, *Tigers are our Brothers: Anthropology of Wildlife Conservation in Northeast India* was published in 2021.

**Sayan Banerjee** is a doctoral researcher at the National Institute of Advanced Studies, Bengaluru, and he is currently examining human–elephant interactions in the state of Assam. He has worked on different projects such as traditional hunting practices and wildlife trade, gender implications of human–wildlife relations, and the nature of local community participation in wildlife conservation projects, all of which are situated in different north-east Indian states. Other than interdisciplinary conservation science, Sayan is also interested in gender, development and environment issues, and plantation geographies.

**Pankaj Joshi** has a PhD degree in Plant Science (Ecology, Taxonomy, and Conservation) from Bhavnagar University, Gujarat. He has over 20 years of research experience in the fields of plant taxonomy, participatory conservation of rare and endangered biodiversity and restoration of natural habitats. He was formerly with the Gujarat Institute of Desert Ecology and has published in a variety of journals. While with Sahjeevan, his work has focused primarily on restoration ecology and resource mapping exercises with the local community.

**Kartik Shanker** is faculty at the Centre for Ecological Sciences, Indian Institute of Science, Bangalore, where he works on the ecology and evolution of frogs, reptiles, birds, plants, and marine fauna. He has set up long-term monitoring programmes for sea turtles across the country. Shanker is also a founding trustee of Dakshin Foundation and founding editor of the magazine, Current Conservation. He is the author of the book *From Soup to Superstar*, a historical account of sea turtle conservation in India, and has written several children's stories.

**Muralidharan Manoharakrishnan** is the Field Director of Dakshin Foundation. His work largely focuses on studying marine flagship species such as sea turtles, sea snakes, sharks, and other similar species across the Indian coast. His work uses widespread marine species to address issues of coastal development and conservation. His primary research interests lie in sea turtle biology, fisheries–wildlife interactions and models of community-based conservation. Having worked closely with traditional fishing communities, Murali has been working towards developing participatory activities in conservation.

**Zai Whitaker** has been interested in natural history and conservation from a young age; she grew up in a family of naturalists and helped launch the India chapter of the World Wildlife Fund. She is the co-founder of the Madras Crocodile Bank and currently is the Managing Trustee. She is the author of 20 children's books about the environment and conservation. These include Kali and the Rat Snake, Andamans Boy and Salim Ali for Schools. Zai was instrumental in starting the Irula Tribal Women's Welfare Society and the Andaman and Nicobar Environment Trust. She is a recipient of the Peter Scott Award for Conservation.

**Kunal Sharma** has worked on issues of conservation, ecotourism, and sustainability in the past. His learnings have revolved around interdisciplinary action research on traditional ecological knowledge of indigenous people, mitigating negative human–wildlife interaction, practical sustainability action, restoration and impacts of tourism formerly with Keystone Foundation, Jungle Lodges and Resorts, GIZ India, and Wildlife Trust of India. He writes in www.livingforest.blogspot.com

**Lokesh Kumar** is an ecotourism professional with more than 18 years of experience with organizations like ECOSS Sikkim, Jungle Lodges & Resorts, Pragya, JPS Associates Consultants, Gujarat Tourism, and the advisory vertical of Yes Bank. His areas of interest include Policy Advocacy, Investment Promotion, Facilitation for State Governments and Skill Development. He has prepared a number of state specific tourism, homestay and skill development policies and worked on numerous internationally funded projects from UNDP, ADB, EU, and JICA among others. His specialization includes Tourism Resource Mapping, Public Private Partnerships, and Techno Commercial Feasibility.

**Madhu Ramnath** has studied about plants in various botanical gardens and herbariums. And he has reoriented his views while living in the forests of Bastar in central India.

# Foreword

Sustainable use is a key component of global efforts to tackle biodiversity loss. It is one of the three foundational pillars of the United Nations Convention on Biological Diversity (CBD), alongside conservation and equitable benefit sharing, and also emphasized as a key conservation incentive in other biodiversity conventions including the Ramsar Convention, the Convention on International Trade in Endangered Species (CITES), and the Convention on Migratory Species (CMS).

Despite this international policy endorsement, the practice of sustainable use is controversial. Many commentators focus on the word *use* and conceptualize sustainable use as the killing of, often iconic, wildlife. But sustainable use takes many different forms – from commercial fisheries to subsistence harvesting of medicinal plants; has many different motivations – to generate income, for cultural and spiritual fulfilment, for recreation; and is carried out by many different stakeholders – governments, private companies, indigenous peoples, and local communities.

The key is ensuring that use – whatever form it takes and whoever does it – is sustainable. That means that it is practised within limits that ensure the long-term survival of the species being used – and hence ensure the ability of those species to continue to provide benefits to people for generations to come.

I am delighted to see this book emerge from the discussions we held in the Nilgiris Hills in India in 2018. Those rich discussions highlighted the multiple different aspects of sustainable use, but also some of the challenges and opportunities. The chapters and case studies presented here provide a wealth of lessons from India that can help inform sustainable use practice across the globe.

This year, a new Global Biodiversity Framework will be agreed on by Parties to the CBD that continues to emphasize the importance of sustainable use. This book provides a timely contribution as we move from agreeing about global policy on sustainable use, to putting it into practice.

Dilys Roe

**Dr. Dilys Roe**
Chair, IUCN Sustainable Use and Livelihoods
Specialist Group (SULi)

# Acknowledgements

The editors would like to acknowledge the funding support from the Global Alliance for Green and Gender Action (GAGGA) received through BothENDS, the Netherlands, for hosting the international workshop in February 2019, where many of these papers were presented, and for subsequent support for the writing and editing of the book.

We would also like to express our sincere gratitude to Madeleine Gefke, an intern with the Keystone Foundation as an Oberlin Shansi Fellow, having graduated from Oberlin College, USA, with majors in Biology and Environmental Studies. Madeleine has been the backbone of our editing and communications for the preparation of this manuscript and played a critical role in bringing this book together.

# 1 An introduction to sustainable use

## And its contribution to biodiversity conservation in India

*Anita Varghese, Mridula Mary Paul, Meera Anna Oommen, and Snehlata Nath*

### Introduction

Sustainable use of biological resources is a polarizing idea in India, both in the literature and in the policy instruments. It is seen by many as an untested and unverified ideal that stands at the opposite end of the spectrum to exclusionary and inviolate protected area conservation – seen as the model that has successfully preserved India's iconic wildlife species and a large share of its biodiversity. Gadgil (1992) wrote about the need for an inclusive model for conservation in India which rests on sustainability and equity and involves participatory planning and management. A decade later, conservation scientists raised questions about the sustainable use paradigm and whether it had room for protection of flagship species (Madhusudan and Shankar Raman 2003).

It remains a question even now, which is why the implementation of the *Scheduled Tribes and Other Traditional Forest Dwellers (Recognition of Forest Rights) Act, 2006*, that gives co-management rights to indigenous people, is highly contested by many in the conservation community. Therein lies the challenge in operationalizing the sustainable use paradigm – while many are convinced of the benefits it will have on the social side, they have serious concerns about its contribution to biodiversity conservation. While elements of sustainable use have been part of the participatory forest management model that was introduced in India in 1990 by the Ministry of Environment Forests Climate Change, it emerged as a distinct conceptual entity in international discussions about conservation in the early 2000s.

The United Nations *Convention on Biological Diversity* defines sustainable use as the 'use of components of biological diversity in a way and at a rate that does not lead to the long-term decline of biological diversity, thereby maintaining its potential to meet the needs and aspirations of present and future generations'. Although sustainable use found a place in IUCN's *World Conservation Strategy* in 1980 when it became part of its mission statement, it was the IUCN *Policy Statement on Sustainable*

DOI: 10.4324/9781003343493-1

*Use of Wild Living Resources* in 2000 that stated that the 'use of wild living resources, if sustainable, is an important conservation tool because the social and economic benefits derived from such use provide incentives for people to conserve them'.

In a similar tone, the Convention on Biological Diversity's 2004 *Addis Ababa Principles and Guidelines for the Sustainable Use of Biodiversity* identifies sustainable use as an effective tool for conservation of biological diversity, and notes its potential for providing key incentives to local communities who derive 'social, cultural, and economic benefits' for conservation. The application of sustainable use as a conservation mechanism is not an easy task and requires effective instruments and institutions of governance that clearly set out incentives and sanctions (IUCN 2000). The imposition of conservation laws in a top-down manner also means that not all communities are in favour of injunctions against use. For a large number of species, this has resulted in their continued extraction, which is carried out illegally. This form of extraction is hugely problematic as access to data and regulation of such practices becomes increasingly difficult. Moreover, illegal harvesting is often associated with unsustainable practices.

## What counts as sustainable use?

What does sustainable use comprise of in the Indian context? At a workshop that was organized in February 2018, key persons from the Sustainable Use and Livelihoods Specialist Group – a joint initiative of the Commission on Environmental Economic and Social Policy and Species Survival Commission of the IUCN – met in Kotagiri, The Nilgiris, Tamil Nadu, with researchers and practitioners of sustainable use across varied Indian landscapes. The meeting was hosted by Keystone Foundation at their office in Kotagiri. Thirty-five participants gave their inputs to draw out lessons from their own work and suggest ways forward on this important issue. The workshop participants listed different activities currently undertaken in forests and oceans in India that could be considered 'sustainable use', attempting to include diverse uses based on varying rules/laws in different regions and states.

Based on the discussions at the workshop, sustainable use in India was divided into consumptive and non-consumptive categories. Consumptive use includes subsistence needs, value addition for household needs, or trade, both small and large scales. Non-consumptive use in India represents cultural, educational, leisure, and sport activities that may range from household to organized commercial scale. 'Consumptive lethal use' would depend entirely on the region/state, if permitted. 'Consumptive non-lethal use' would depend on whether the rules of harvest are followed or destructive methods of harvest are kept in check.

*Table 1.1* List of sustainable use practices and categorization for India

| | *Consumptive* | | *Non-consumptive* |
|---|---|---|---|
| | *Lethal* | *Non-Lethal* | |
| **Subsistence** | Medicinal and aromatic plants (MAPs) and plant non-timber forest products (NTFPs)<br>Seaweed | MAPs and plant NTFPs<br>Seed collection for propagation | Sacred groves<br>Sacred species |
| **Commercial** | MAPs and plant NTFPs trade<br>Seashell collection | Medicinal plant trade<br>Teeth and feathers | Nature tourism<br>Diving |
| **Recreational** | Fishing<br>Recreational hunting | Angling<br>Pets | Nature tourism<br>Diving |

## How this book is structured

The chapters in this book were selected in an attempt to have at least one representative case study from each of the categories listed in Table 1.1. Efforts were made to get in touch with key persons in varying biogeographical regions of India so that at least one representative case study from each of these regions is included in the collection. The authors are drawn from academia and practice and are associated with NGOs working in the field. Some of these are personal accounts, and we have been mindful to incorporate narrative styles outside the strictures of academia. Forest managers and conservation policymakers were invited to be a part of the writing. However, owing to constraints on time, this was not to be. We aim to address this limitation in future editions, in order to make the book more inclusive in terms of the perspectives it brings together.

Based on the learnings from the workshop and subsequent case studies, the book is structured along the following themes – governance, enterprises, community knowledge, and intangible benefits.

The next chapter by the editors seeks to situate the sustainable use paradigm within the Indian context, bringing out its significance in policy, application in the lives of people for whom it is an important part of livelihood, current good/positive practices, and contribution to biodiversity conservation.

Thereafter, Mridula Mary Paul explores the historical legacy and provisions in contemporary Indian policy for conservation and people's rights to access natural areas. The trajectory of governance around sustainable use in India is discussed at length in the paper through the legislations that are in place and how this will aid in operationalizing sustainable use as a conservation tool in the country.

Naveen Namboothri et al. discuss the case of fishers in the oceanic atolls of Lakshadweep islands and how transitioning from cultural practices, pressures from developmental activities, differences between state and local management, etc., pose a distinct challenge to continuing sustainable practices in traditional fishing activities.

Snehlata Nath looks at the facets of commercial NTFP harvest that hold potential for sustainable use through value addition, ecological monitoring, or ex situ propagation. This is done through the experiences of field NGOs that have been working in the area for more than 25 years. The lessons learned and challenges encountered are highlighted in the chapter.

Jagannatha Rao R. et al., through their long-term work on sustainable harvest practices in medicinal plant collection from the wild, describe their working experience with the state machinery and the collaboration that was brought about to ensure sustainable harvesting practices.

Meera Anna Oommen discusses the traditional hunting practices of islanders in the Andamans to highlight how social taboos and cultural rules surrounding the practice play an important role in keeping the resource management sustainable. For the recent settlers, the traditional practices do not hold significance, which leads to opportunistic and unsustainable harvests.

Ambika Aiyyadurai et al. explore the idea of social and human dimensions of hunting as practised among the communities in parts of north-east India. The indigenous worldviews, gender roles, governance, and cultural facets bring to focus how hunting is not just about trade and consumption.

Pankaj Joshi writes about the grazing practices of the indigenous people of the Banni grasslands in Gujarat. The community has traditionally practised a sustainable form of managing the grasslands and in documenting them, there is potential to look through the lens of sustainable management at other landscapes as well. Hunting and grazing are too often regarded as a taboo topic for natural resource management.

Kartik Shanker and Muralidharan Manoharakrishnan share insights from a long-term citizen science effort that engages civil society and other role players to monitor the late maturing, long living, migratory marine turtles along the eastern coastline of India in Chennai and Odisha. These participatory monitoring efforts have given rise to research and conservation efforts at a national level.

Zai Whittaker shares the knowledge and experiences of the famous reptile conservation efforts of the Madras Crocodile Bank, from how it was set up and what lessons were learned along the way. Education played a major role in influencing how sustainable use could become a part of conservation.

Kunal Sharma et al. discuss the efforts of state-run and private efforts at nature tourism. While nature or ecotourism has the potential to be a high impact of low footprint effort that could be a win for livelihoods and conservation, it has grown in India over the past decades without many checks and balances. In tracing the history of the industry, the authors bring out opportunities and challenges for the road ahead.

Madhu Ramnath discusses the role of sacred groves in the cultural and social fabric of indigenous people of Central India. The traditions are set deeply and the groves influence human behaviour to become part of protection and conservation.

The editors conclude this volume by highlighting the lessons learned from the several ongoing efforts in the country with respect to the sustainable use of nature in relation to inclusive conservation. The varied experiences of researchers, practitioners, and community members bring to the forefront challenges that lie ahead in making sustainable use a conservation goal within the context of India. The work over several decades provides many lessons for the future. Sustainable use in conservation requires much work on several fronts to keep the momentum while adapting to newer challenges.

## References

Gadgil, M. (1992). Conserving biodiversity as if people matter: A case study from India. *Ambio*, *21*(3), 266–270. www.jstor.org/stable/4313937

IUCN. (2000). *The IUCN policy statement on sustainable use of wild living resources adopted at the IUCN World Conservation Congress, Amman*. https://portals.iucn.org/library/efiles/documents/Rep-2000-054.pdf

Madhusudan, M. D., & Shankar Raman, T. R. (2003). Conservation as if biological diversity matters: Preservation versus sustainable use in India. *Conservation and Society*, *1*(1), 49–59. www.conservationandsociety.org.in//text.asp?2003/1/1/49/49356

# 2 Sustainable use and biodiversity conservation

## When the twain shall meet!

*Snehlata Nath, Meera Anna Oommen, Mridula Mary Paul, and Anita Varghese*

### Introduction

Traditional ways of exclusionary conservation or management of protected areas view people's dependence on wild resources as a threat to biodiversity. This point of view also tends to regard 'use' as solely extractive and the root cause of resource depletion. However, interconnectedness between people and nature has always existed in landscapes where human civilizations have thrived. The Food and Agriculture Organization's State of the Forests report for 2020 rightly identifies agriculture expansion as practised in commercial cattle ranching, and soy and palm oil plantations as the main driver for forest loss. This makes it clear that not all types of agriculture (especially not the food-growing subsistence type) have the same impact.

The report goes on to state that 880 million people are dependent on forest-based activities such as charcoal production and collection of fuelwood, and that many low-income countries have high forest cover and biodiversity richness with low human density and high rates of poverty (FAO 2020). In such cases, it is difficult to make direct connections between poverty and biodiversity without an understanding of underlying factors. Barrett et al. (2011) showed, through a collection of research findings from around the world, that factors such as dependence on limited natural resources, shared vulnerabilities, failure of social institutions, and lack of informed adaptive management play a large role in keeping biodiversity conservation and poverty traps interlinked.

### Legacy of use

In megadiverse regions such as the Western Ghats of India, human population density is very high, which can pose a threat to biodiversity conservation efforts (Cincotta et al. 2000). Yet, we also know that such landscapes continue to support high levels of endemics and viable populations of endangered charismatic megafauna such as tigers and elephants. It is the paradox of increasing anthropogenic pressures, while maintaining relatively good biodiversity richness that makes several parts of India's forests an enigma.

DOI: 10.4324/9781003343493-2

This leads us to pose larger questions – what types of human use are less detrimental to biodiversity? Which areas of biodiversity support human wellbeing? Are the roles of human use and biodiversity conservation complementary? While all three are important questions to reach an understanding of the role of sustainable use in biodiversity conservation, the scope of this chapter rests on exploring the third question on the complementarity of human use and biodiversity conservation.

The understanding that certain conditions of biodiversity richness can go hand in hand with human well being is increasing, especially under the 'sustainable use' paradigm. Though in some parts of the world, discussions about sustainable use tend to focus on terrestrial wildlife and trophy hunting, in India sustainable use largely means harvesting and trading non-timber forest products (NTFPs) including honey and medicinal plants, which are collected from mountains, forests, and oceans. Marginalized and impoverished communities are usually the harvesters of these NTFPs and they are faced with uncertain tenure (especially in protected areas) and poor monetary returns.

On account of a number of factors, traditional knowledge regarding natural produce is also eroding. Women play an important role in harvesting of wild resources and are very often overlooked when decisions are made about the economics and ecology of products derived from these natural systems. The varied components of biodiversity richness and human wellbeing need to be considered when addressing their interconnectedness.

India has a long history of use of wild flora and fauna. Over the millennia, communities have reaped nutritional, economic, cultural, and spiritual benefits, largely guided by traditional systems of use and governance. Historical evidence for multifaceted use can be found in the Vedic pharmacology that considered the 'universe as a kitchen' and ancient Indian zoology and botany that often found representation as a 'catalogue of meats' or in the language of food and medicines (Zimmerman 1982).

As a megadiverse region with over 17,000 species of higher plants and substantial diversity and endemism in other taxa, the use of wild resources is considerable, and there remain untapped resources which can further benefit human societies. The Indian subcontinent is believed to be the source of a large number of medicinal resources and also the centre of domestication for over 30% of crop species, many of which are economically valuable. Many higher order taxa ranging from reptiles and amphibians to birds and mammals find a place in the spectrum of use by human communities.

While plant and lower-order resources continue to be extracted (and in many cases exploited extensively), contemporary India has witnessed shifts in traditional systems of utilization of wild animals. Hunting or harvesting of large game which was commonplace across the region is no longer allowed except for traditional societies in select geographies such as northeast India and the Andaman and Nicobar islands. In many areas, the use of wildlife is entirely prohibited, bringing an end to centuries of these interactions between local communities, forests, and animals.

## Prohibiting use

These developments reflect two key drivers of policy in India. First, in light of threats to biodiversity and ecosystem services, the positioning of wildlife and plants as renewable resources began to be questioned by scientists, managers, and conservation agencies on the basis of long-term sustainability. Second, much of this questioning stemmed from elite views and values with regard to wildlife, starting with princely and colonial game laws, which labelled subsistence use by local communities as detrimental to sport, and followed by post-Independence and global protectionist enterprises such as fortress conservation, which have received support from the state and elite conservation groups in India.

The injunction against use was sealed when the proposed *Wildlife Conservation and Management Bill 1972* was reframed by the then Prime Minister, Mrs. Indira Gandhi, as the *Wildlife (Protection) Bill* with the explicit intention of avoiding the 'killing' of animals (Ramesh 2017). This Bill was passed by Parliament, becoming the *Wildlife (Protection) Act, 1972*, India's landmark legislation that favours elite ideologies of conservation, while giving short shrift to centuries of sustainable use of nature by local communities.

Under this law, the hunting and harvesting of most species of wildlife, excluding a handful that has been classified as 'vermin', are prohibited. Such bans entail the loss of usufruct rights, livelihoods, and incomes. In this context, it needs to be emphasized that long-term histories of India dispel the myth of a pristine past. Forests served as sites of use, although not always of a sustainable nature (Morrison 2014). While sustainability of natural resource use needs to be addressed in light of ongoing and emerging challenges, a complete break from natural resource-based livelihoods was catastrophic.

For instance, in 2001, a sudden ban effected by the Ministry of Environment Forest and Climate Change on elasmobranch fisheries (comprising of sharks, rays, skates, etc.) in India threw the shark fishing communities into sudden disarray. Eventually, opposition to this top-down conservation measure resulted in a shift to selective restrictions rather than a complete blanket ban on shark fishing. Similarly, a blanket ban on removal of all NTFPs from wildlife sanctuaries in 2004 significantly impacted the lives of indigenous Soliga people, preventing them from entering the forests that were a part of their ancestral domains for centuries (Madegowda and Rao 2013).

For communities that are marginalized or living outside mainstream livelihood systems, restrictions on use mean that a recourse to wild foods is no longer an option, leaving the door open for nutritional deprivation, especially of protein. Long-term food insecurity impacts their overall well-being. As pointed out within anthropological scholarship, forest-dwelling communities often incorporate an astounding diversity of species into their diets. For instance, Ramnath (2015) points to the careful and sustained use of over 300 food plant species by the Durwa people of Bastar. The Durwa supplement plant foods with wild game and fish. Knowledge about species, their habitats, and seasonality are important to enable use, and if employed for management, it could very well be used for conservation.

Since subsistence use systems are typically multifaceted, the clamping down on natural resource use as a result of these blanket bans also results in the loss of traditional knowledge and cultural practices associated with such systems. The loss of traditional practices is often accompanied by a loss of intergenerational knowledge about sustainable management practices and ecological resilience. Gentrification attempts (e.g. the Pardhis in Madhya Pradesh) (WWF 2010) aimed at tribal communities with an occupational history of traditional hunting is a case in point of conservationists ignoring wider traditional and cultural systems of knowledge and practice that are in danger of being lost in favour of narrow protectionist gains.

## When use complements conservation

Rather than using restrictive bans for conservation, the case of the Andaman Islands, where hunting by local communities is allowed, illustrates that cultural practices are often tied to ecological sustainability. Communities such as the Ongee and the Jarawa impose restrictions and taboos on hunting and consumption of wild boar and sea turtles as a part of key individual milestones such as births and deaths, coming of age, and initiation rituals. Pigs and turtles are hunted at designated places, during particular seasons, and within specific territories. These practices are tied up within these communities' cosmology and creation myths (Man 1883; Pandya 2009). Colding and Folke (2001) refer to such resource and habitat taboos (RHT) as 'invisible systems' that play significant functional roles in the long-term management of populations and also provide benefits for conservation (see also Reichel-Dolmatoff 1976).

The maintenance of 'biocultural systems' that uphold sustainable and resilient interactions between people and nature are critical to conservation (Maffi and Woodley 2010). Many communities in India practice sustainable use along these lines. The wild honey collections done in the forests of the Nilgiri Mountains in the Western Ghats present a rich cultural heritage where the tribal honey hunter undertakes this life-threatening practice year after year (Roy et al. 2011). To the honey hunter it is business as usual and an inherent part of their identity and culture to which they must return even if they are employed elsewhere – coming home to the village in the honey season is a necessity. The migratory bees return to the mountains every year and bee populations have been found extant in these landscapes (Varghese et al. 2015).

In the case of the Soliga of Biligiri Rangan Temple tiger reserve in Karnataka, long-term ecological studies in the forests where they collect NTFP have shown that the ban on *amla* collection imposed in 2004 did not help to protect the *amla* trees. Research showed that even though fruit harvesting can significantly decrease population growth rates of the *amla* trees, the effect of fruit harvest was much lower than that of hemi-parasitic plants like mistletoe and invasives like *Lantana camara* (Ticktin 2015).

Disproportionate attention on charismatic species is yet another feature of Indian conservation that has consequences for sustainable use. While some species, especially large mammals get more attention, most other groups including small mammals, birds, reptiles, amphibians, and fishes are currently not adequately scrutinized. Plants do receive some attention, though trees may garner a greater share of public interest when compared to shrubs, climbers, and grasses. Sometimes attention comes about when a species is facing extinction, as in the case of the caterpillar fungus *Cordyceps sinensis* found in the Himalayan ranges, threatened because of over exploitation. Conservation funding, outreach, action, and policy seem directed at species and ecosystems which are under threat, and this reduces the opportunity to work on species that are used and about which local communities have knowledge.

## An indifferent conservation regime

India has a maze of conservation regimes, and ecologically rich areas are accordingly categorized into different types of protected areas, managed by state or national entities, with varied legislative mechanisms that set out the extent of protection and use possible within them. Policies and regulations relevant to conservation and sustainable use are contained not only in environmental legislation but also in broader developmental policies that are formulated by a range of government agencies.

The *Indian Forest Act, 1927,* contained many restrictions on use of natural resources and practices such as grazing, hunting, and collection of forest produce. The *National Forest Policy of 1952* continued in the same spirit, though with some contradictions, whereby linkages were made to local communities, their livelihoods, and forest management. The *Wildlife Protection Act, 1972*, restrictive as it was, did provide some access to wildlife sanctuaries for nearby communities. The *National Forest Policy of 1988* was perhaps the first legislation that spoke boldly for sustainable use and rights of communities over forest produce with due consideration to the carrying capacity of the area. From 1990 onwards, the Joint Forest Management policies of the Government started to proactively advocate for benefit sharing with communities and the co-management between government and local communities of NTFPs, until then referred to only as minor forest produce.

Several development policies from the late 1990s specified the need to see protection and conservation of the environment as an important part of sustainable use of resources. India's commitments to the Convention on Biological Diversity (CBD) saw the formulation of the National Biodiversity Act of 2002 adopting the CBD definition of sustainable use, which although broad, does not specify access for use by communities.

In a certain departure from these tepid affirmations of sustainable use, the *Scheduled Tribes and Other Traditional Forest Dwellers (Recognition of Forest Rights) Act* (FRA) enacted in 2006, unequivocally recognized local communities as custodians of sustainable use and conservation of biodiversity, whose practices further the conservation regime of forests. This legislation has further divided the conservation community in India, with some feeling strongly that this will undermine conservation efforts taken after India's independence. Others contend that the conservation practices thus far have not been fair to India's indigenous people and that this Act brings justice to these communities and holds a larger potential for more just and equitable form of conservation. A detailed analysis of the policy environment regarding sustainable use is presented in the next chapter.

There are several biodiversity rich areas where local communities are better off in terms of food security and nutrition in comparison with their relatives who have been relocated to areas with better access to roads and towns. Several protected areas in the Western Ghats in India have highly degraded habitats, and reduction of human pressure through relocations has not arrested habitat degradation especially with regard to invasive species. We need improved understanding and communication around successful systems that serve both human well-being and biodiversity richness. This is of crucial importance before we simply refer back to principles of 'fortress' approaches, whether in human development or in conservation, that take us further away from the progress we have made globally with interdisciplinary and cross-scale approaches so far.

## Conclusion

In the last two decades, protected area coverage in India has increased and so has protection for indigenous people's access and tenure rights. Even so, the overall quality of habitats and human life is not improving. Increased human–wildlife conflicts, widespread colonization of invasive species, loss of grasslands, lowered water tables, rising ocean temperatures, and prolonged periods of drought persist and pose challenges to biodiversity and natural resource use. Marginalized indigenous communities, who live by subsistence use, are most threatened by these challenges. We anticipate that climate change will further exacerbate some of these threats, reducing the resilience of the environment and increasing the vulnerability of marginalized people.

Our observations through research, practice, and interaction have revealed that communities' life and nature interactions have sustainability science built into traditional methods, which is crucial for maintaining and enhancing the quality of their environment. These adaptive strategies embedded in socio-ecological practices draw from culture, landscapes, beliefs, and experience. While market signals and administrative rules of the State can disrupt feedbacks that structure adaptive management of socio-ecological systems, we see the potential to harmonize these differently scaled resources for improved environmental governance. The effects of exclusion of traditional use practices and management of forest resources are unclear and need to be monitored closely to avoid both biodiversity loss and negative impacts on local livelihoods.

## References


Cincotta, R. P., Wisnewski, J., & Engelman, R. (2000). Human population in the biodiversity hotspots. *Nature*, *404*(6781), 990–992.

Colding, J., & Folke, C. (2001). Social taboos: 'Invisible' systems of local resource management and biological conservation. *Ecological Applications*, *11*(2), 584–600. https://doi.org/10.1890/1051-0761(2001)011[0584:STISOL]2.0.CO;2

Madegowda, C., & Rao, C. U. (2013). The ban of non-timber forest products collection effect on Soligas migration in Biligiri Rangaswamy temple wildlife sanctuary, India. *Antrocom Online Journal of Anthropology*, *9*(1), 105–114.

Maffi, L., & Woodley, E. (2010). *Biocultural diversity conservation: A global sourcebook*. London: Earthscan. DOI:10.14237/ebl.3.2012.46

Man, E. H. (1883). On the aboriginal inhabitants of the Andaman Islands. *The Journal of the Anthropological Institute of Great Britain and Ireland*, *12*, 69–175. https://doi.org/10.2307/2841843, https://doi.org/10.2307/2841953

Morrison, K. (2014). Conceiving ecology and stopping the clock: Narratives of balance, loss and degradation. In M. Rangarajan & K. Sivaramakrishnan (Eds.), *Shifting ground: People, animals and mobility in India's environmental history* (pp. 39–64). Oxford: Oxford University Press. DOI:10.1093/acprof:oso/9780198098959.003.0002

Pandya, V. (2009). *In the forest: Visual and material worlds of Andamanese history (1858–2006)*. Lanham, MD: University Press of America. DOI:10.1017/S0021911810002792

Ramesh, J. (2017). *Indira Gandhi: A life in nature*. Noida: Simon & Schuster.

Ramnath, M. (2015). *Woodsmoke and leafcups: Autobiographical footnotes to the anthropology of the Durwa*. New Delhi: HarperCollins Publishers India.

Reichel-Dolmatoff, G. (1976). Cosmology as ecological analysis. *Man*, *11*(3), 307–318.

Roy, P., Leo, R., Thomas, S. G., Varghese, A., Sharma, K., Prasad, S., . . . & Davidar, P. (2011). Nesting requirements of the rock bee Apis dorsata in the Nilgiri Biosphere Reserve, India. *Tropical Ecology*, *52*(3), 285–291.

Ticktin, T. (2015). The ecological sustainability of non-timber forest product harvest: Principles and methods. In *Ecological sustainability for non-timber forest products* (pp. 45–66). Routledge.

Varghese, A., Nath, S., Leo, R., & Thomas, S. G. (2015). The road to sustainable harvests in wild honey collection. In C. M. Shackleton, A. K. Pandey, & T. Ticktin (Eds.), *Ecological sustainability for non-timber forest products: Dynamics and case studies of harvesting* (pp. 103–115). London: Routledge.

World Wildlife Fund (WWF). (2010, April 22). *Change through education*. www.wwfindia.org/?4301/Change-through-education#:~:text=A%20reformed%20life%20for%20children,creative%20engagement%20and%20formal%20education.

Zimmermann, F. (1982). *The jungle and the aroma of meats: An ecological theme in Hindu medicine*. Delhi: Motilal Banarsidass Publishers. DOI: 10.1016/0277–9536(88)90121–9

# Part I

# Governance

# 3 The governance of sustainable use

## Historical legacy and contemporary deployment

*Mridula Mary Paul*

### Introduction

In a country like India, with its biologically diverse landscapes and a vast array of communities and cultures, the 'wild' has always been lived spaces. People's use of wild resources, consequently, was a commonplace part of life and such use was widespread (Guha 2002). Ten thousand years ago, before the distinction between the wild and civilized were delineated on a map, people hunted and collected produce from the lands around them (Rangarajan 2007). 5,000 years following that, we have evidence of herders in the Vindhyas using wooden structures to corral their sheep (Rangarajan 2007).

Power has always had a part to play in resource use (Guha 2002). Governing access to it is a political act that manifested in the form of mores, norms, and laws at different points in time. Through a legal review that places itself in the context of the environmental history literature, this chapter aims to inject the political into sustainable use (SU) in the Indian context.

In the following sections, sustainable use is used as a catch-all phrase to capture all methods and manner of use and access to natural resources located outside agricultural areas and human settlements. The Convention on Biological Diversity (CBD) provides a technical definition of sustainable use that covers a rate and manner of use that does not compromise biological diversity or continued future utilization. Described as an 'umbrella term', it does not solely comprise consumptive use or restrict such use to local communities (Hutton and Leader-Williams 2003).

However, given that access to wild resources has enormous implications for the lives and livelihoods of people (Rangarajan 2006), this chapter focuses on the sustainable use of such resources by local communities. The sections that follow will trace the historical and contemporary trajectory of governance around sustainable use in India and explore the implications of this for operationalizing sustainable use as a conservation tool in the country.

DOI: 10.4324/9781003343493-4

## Early governance of SU

Early Indian kingdoms had strictures in place governing the use of wild resources. Kautilya's Arthasasthra contains a proto-version of the doctrine of eminent domain that asserted state authority over all wild resources (Thapar 2007). In pre-colonial times, access to these resources from the wild was loosely regulated, except in the case of certain species or specific areas (Rangarajan 2007; Rangarajan 1996; Guha 2002). The third BCE Mauryan kingdom claimed state monopoly over elephants and invoked strict penalties against the poaching of these animals (Rangarajan 2007). In the 1800s, certain grasslands were reserved for the fodder needs of Maharashtra's ruling armies (Guha 2007; Guha 2002).

Underlying these governance norms were economic considerations, particularly in relation to resources such as timber and gemstones that had become valued commodities (Thapar 2007). Governance was then a way that the powerful appropriated use for themselves through the exclusion or dilution of local use. The need for such governance was justified in the context of the emerging narrative that equated civilization with the clearing of jungles and derided a forest-dependent lifestyle (Rangarajan 1996; Thapar 2007). However, because many pre-colonial kingdoms also depended on forest-dwelling or nomadic peoples as archers or herders of pack animals used to transport supplies during warfare, the relationship between the state and local resource users was, even if rife with contestations (Guha 2007), not always antagonistic (Rangarajan 1996).

## SU in the colonial state

Although control of resource use by the powerful, through the exclusion of others, predated the colonial state (Guha 2002), it is during the colonial regime that the allowances for local resource use were considerably whittled down (Skaria 1992). The position of forest dwellers changed as colonial rule progressed, and from being seen as allies in the process of clearing 'wild' jungle for cultivation, they soon began to be viewed as competitors for game which were to be the sole preserves of the British elite, whether for hunting or for preservation (Rangarajan 1998).

Yet even within the colonial administration, there were considerable disagreements regarding the role of local resource users. The Madras Government was opposed to the Draft Forest Bill circulated in 1869, pointing out that such a restrictive and overarching forest legislation had no precedent in Indian law and that it would have serious detrimental impacts on the livelihood of forest-dependent communities (Guha 2001). The Act was nevertheless passed in 1878, but local resource users had an unlikely source of support in Dietrich Brandis, the first Inspector General of Forests, who wrote a series of reports exhorting the colonial forest department to establish village forests for the use of communities dependent on the forest, with the stated motive that doing so would ensure that they would be invested in the protection of forests (Guha 2001).

These suggestions did not sit well with the colonial state which was of the view that 'natives' were irresponsible users of wild resources (Guha 2001; Rangarajan 1996). Ironically, the colonial project, with its philosophical underpinnings located in the conquest of nature (Thapar 2007), is not considered similarly profligate, even when it involved large-scale felling of timber for the railways or mining for coal. Resource use in this context was imbued with a higher moral purpose, despite the widespread ecological destruction that attended it (Lal 2007), the effects of which continue into contemporary times (Rangarajan 2007). Therefore, the question of the use of wild resources has broader political and economic imperatives underpinning it (Guha 2007). At it's heart lies decisions regarding who constitutes a threat to the wild (Rangarajan 2007), and consequently what kind of use is acceptable.

Restrictions on resource use practices, such as nomadic pastoralism, were justified by drawing on the prevalent scientific opinion of the time that linked grazing with desertification and denudation (Agrawal and Saberwal 2007). Similar arguments were made in the case of swidden cultivation (Agrawal and Saberwal 2007). Around the same time, state-built canal systems on the Indus and the Ganges were creating salt deposits and causing widespread loss of soil fertility (Rangarajan 2007), but received no censure. The state was, therefore, the arbiter of what manner of environmental change was deemed degradation. The vilification of local resource use was a strategically adopted narrative that foregrounded policies that allowed for blanket state control over wild resources. The *Indian Forest Act*, which came into effect in 1927, put this into operation through the categorization of forests into those that were 'reserved' and 'protected', among other variations. Appended to these classifications were prohibitions and restrictions to varying degrees, of rights of use within them, including the collection of forest produce, grazing, hunting, and shifting cultivation.

## SU principles in contemporary policies

The *National Forest Policy, 1952* was among independent India's early environmental policies. It declared forests as assets held by the state in public interest and although it acknowledged these as 'irksome', restrictions on local usage were deemed necessary in the interests of national well-being. Somewhat contradictorily, the Policy also noted that the livelihood linkages between forests and local communities are beneficial for forest management and highlighted the importance of local support that hinged on making provisions for their 'direct interest in the utilization of forests'.

This viewpoint did not, however, translate into operational provisions, and instead additional restrictions and prohibitions to use were introduced by the *Wildlife (Protection) Act, 1972,* which established further categories of protected areas in the form of 'sanctuary' and 'national park'. This Act provided for some use, for instance – allowing for the collection of forest produce for the personal use of communities living near sanctuaries. Grazing was allowed in sanctuaries, subject to regulation, although not in national parks. The new conservation regime with stricter norms on resource use involved the commutation of rights in protected areas and the removal of villages to locations outside such areas (Rangarajan 1998).

At the international stage, however, the discourse around the sustainable use of species and ecosystems as a conservation strategy was taking shape and in 1980 found a place in the mission statement of IUCN's World Conservation Strategy. The *National Forest Policy, 1988* drafted in this context departed from the earlier forest policy when it referred to sustainable use as a facet of conservation, noting that the communities living near forests must be treated as having the primary right over forest produce, within the limits of the forests' carrying capacity. The *Joint Forest Management Guidelines, 1990* (JFM), which this Policy helped spawn, advocated benefit-sharing with communities.

JFM was an ambitious venture of the Indian forest administration to involve local people in forest management through formalizing access to resources. It created committees of forest-dependent populations conscripted for the protection of forests and forest resources, in return for which services they were entitled to certain benefits such as a share in the timber harvested (Sundar 2000). JFM was formulated in response to various protest movements and resource conflicts prevalent in many developing countries against coercive forest governance regimes, accompanied by much debate in academic and scientific circles (Guha 2001; Rajan 2006; Sundar 2000). In a departure from the colonial project of vilifying the destructive native, it embraced a narrative that hailed the virtues of the 'primitive' conservation practices of the forest dweller (Agrawal and Gibson 1999) and alluded to a 'lost tradition' of sustainable resource use (Mosse 1997, p. 257).

In keeping with this theme and in context of the 1992 Rio Earth Summit, India's central planning body, the Planning Commission, specified in its *8th Five Year Plan (1992–97)* that the protection and conservation of the environment is necessarily linked to the sustainable use of resources, and mandated that plans for environmental management include provisions for such use. Along similar lines, the *National Conservation Strategy and Policy Statement on Environment and Development, 1992* noted that the country's development challenges could be met by securing the fundamental needs of people through sustainable and equitable use of natural resources, although it largely focused on such use in common lands and degraded forests. By 2000, the IUCN *Policy Statement on Sustainable Use of Wild Living Resources* had articulated that the 'use of wild living resources, if sustainable, is an important conservation tool because the social and economic benefits derived from such use provide incentives for people to conserve them'.

In 2002, the *National Biodiversity Act*, enacted in fulfilment of India's commitments under the CBD, directed the national government to develop strategies and programmes for the sustainable use of biological diversity and integrate such use into existing government plans. This Act, which largely reproduces the CBD definition of sustainable use, is a predominantly procedural one, concerned with licenses and rules of access for research and commercial use. It speaks little to sustainable use of natural resources by local communities and the opportunities these pose to strengthen conservation of those resources. CBD's 2004 *Addis Ababa Principles and Guidelines for the Sustainable Use of Biodiversity* that sets out sustainable use as an effective tool for conservation of biological diversity, providing as it does key incentives for conservation for people, who derive 'social, cultural, and economic benefits' from such use, find no mention.

The Planning Commission's *10th Five Year Plan (2002–07)* reiterated the strong linkages between conservation and sustainable use, particularly in rural areas, and put sustainable use of biological resources at the core of ecologically sustainable development. The sustainable use paradigm received support from the courts, with the Supreme Court in 2005 recognizing that the sustainable use of biological resources is fundamental to ecological sustainability in *T.N. Godavarman Thirumulpad vs Union of India.*

Furthermore, fillip came from the *National Environment Policy, 2006* which adopted as its dominant theme, the rationale that the most secure basis for conservation lies with ensuring that local communities better their lives through it. It noted that development policies that unwittingly prioritized the individual over the community had sounded the death knell for traditional institutions that governed the sustainable use of natural resources, with adverse implications for the environment. It placed the blame on disincentives for conservation, on institutional failure that manifests in unclear or insufficiently enforced rights of access to, and use of, environmental resources.

In something of a watershed moment, the *Scheduled Tribes and Other Traditional Forest Dwellers (Recognition of Forest Rights) Act* (FRA) enacted in 2006, unequivocally recognized local communities as custodians of sustainable use and conservation of biodiversity, whose practices further the conservation regime of forests. It referred to the rich conservation history of these communities enshrined in their traditional, sustainable methods of natural resource use, and granted rights of access and use of natural resources, subject to the processes prescribed under the Act. The resource use prescribed included the collection of minor forest produce, grazing, and fishing, but specifically excluded hunting.

The year of 2007 saw the formulation of the *Protection, Development, Maintenance, and Research in Biosphere Reserves in India Guidelines* in compliance with UNESCO's Man and Biosphere Programme that focused on landscape-level conservation in and around existing protected areas. Among its list of functions related to conservation, it specified the promotion of traditional and sustainable systems of resource use. Nevertheless, it spatially separated such use, confining it to 'buffer areas', while declaring 'core zones' that were to be devoid of any human pressure. The year 2007 also saw the Planning Commission pointing out the interconnectedness of conservation and sustainable use, calling for innovations to sustain economic growth that can meet developmental objectives, in conjunction with sustainable use of natural resources in its *11th Five Year Plan (2007–12)*.

The final *12th Five Year Plan (2012–17)* of the Planning Commission associated sustainable use of biodiversity with livelihood security and highlighted the importance of this, particularly in ecologically sensitive areas such as the coastal zone, Western Ghats, and other hilly areas. In a landmark judgement supportive of this view, the Supreme Court in *Orissa Mining Corporation Ltd vs. Ministry of Environment & Forest* (2013) held that sustainable use by forest-dwelling communities would boost the tenets of conservation within forests, while ensuring livelihood security for these communities.

In contrast to the position taken by the Planning Commission, its successor institution, the NITI Aayog in its *Three Year Action Agenda (2017–18 to 2019–20)*, makes no mention of sustainable use of natural resources, except for ground water and the removal of restrictions around harvesting trees in private lands. There is some focus on the sustainable use of natural resources by the private sector, but the Agenda makes no reference to use by local communities. Sustainable use does, however, feature prominently in India's *National Biodiversity Targets* under CBD (MoEFCC 2014).

## Departure from SU principles in recent policies

The *Third National Wildlife Action Plan (2017–30)* is an exercise in contradiction. It notes the need to bring the current management of protected areas up to speed by accommodating the 'genuine needs' of local populations, but couches this in a narrative of not causing further 'alienation of local people'. It refers to the amendments to the *Wildlife (Protection) Act, 1972* introduced in later years creating further layers of protection, such as through the establishment of 'tiger reserves', 'conservation reserves', and 'community reserves', and clearly states that these newer modes of protection have been necessitated by the difficulties in constituting full-fledged protected areas such as national parks and sanctuaries.

At the same time, it lists as a mandate the establishment of new national parks and sanctuaries, where sustainable use is statutorily curtailed. It also looks to create wildlife corridors that can extend the network of tiger reserves. In justification, the Plan notes that resource use, even at subsistence levels, puts pressure on protected areas, and implicitly therefore, on conservation. This Action Plan sees the Forest Rights Act as diluting the sanctity of protected areas, attributing to it a likely escalation in human–wildlife conflicts, intensification of human pressure, and the undermining of the functioning of JFM institutions. While it acknowledges that public support is crucial to wildlife conservation and sees the benefits of ensuring that locals meet their livelihood needs from protected areas, it seeks to direct these initiatives through institutions that are familiar to the forest bureaucracy such as van panchayats, JFM committees, village forests, and eco-development committees. In doing so, it looks to distance conservation from the language of rights contained within FRA. A draft Wildlife Policy is in the process of being deliberated in closed government fora (Lok Sabha 2020).

Introduced in 2018, the draft Forest Policy speaks about improving livelihoods based on sustainable use of ecosystem services and refers to minor forest produce as an important source of livelihood for forest-dependent communities. It looks to set up a National Community Forest Management Mission to provide an operational framework for participatory forest management that builds on existing JFM bodies. This Policy is fairly clear in its emphasis that biodiversity conservation requires protected areas, which must be extended, with strengthened enforcement in existing ones. While there are broad references to managing forest produce sustainably, there is little focus on issues of use or access by local communities.

In March 2019, a comprehensive amendment was proposed to the *Indian Forest Act, 1927,* that reveals a modified preamble that is cognizant of the well-being of people, particularly forest-dependent people, in the context of forest conservation. The proposed Act refers to a landscape approach to conservation in certain areas, which will address matters of conservation, sustainable use, and production, and grants powers to state governments to regulate sustainable use in forests that are not government-owned. In a contradiction to its own preamble, the proposed Act seeks to dilute, and where feasible, extinguish rights of access and use that have been granted to local communities by the FRA as part of provisions relating to establishing new protected areas. It also provides for increasing penalties for violations and assigns greater punitive authority to the forest bureaucracy. This amendment was withdrawn in November 2019 (PIB 2019), having received widespread criticism from civil society.

The 2017 *National Wildlife Action Plan* had called for a review of the *Wildlife (Protection) Act, 1972* in order to incorporate provisions for the equitable sharing of the benefits of conservation and ecosystem services between government agencies and local communities, with an eye to complying with India's international obligations. The *Wildlife (Protection) Amendment Bill,* 2021 that followed does incorporate provisions to meet India's commitments under international conventions, but appears reluctant to address the first part of Action Plan's proposition. It does, however, make a marginal nod to FRA by adding a provision that requires the Chief Wildlife Warden to collaborate with the Gram Sabha to develop management plans in sanctuaries. It also makes concessions for legitimate household water requirements by locals. Introduced during the same period, the *Biological Diversity (Amendment) Bill, 2021* makes no attempt to redress the limited references to sustainable local use in the original Act, preferring instead to ease processes for commercial use by corporations.

## Lessons from the deployment of SU

As these instruments reveal, India has no dearth of laws, judicial pronouncements, and policies around sustainable use. The range of policies outlined earlier makes the appropriate noises about sustainable use and its importance for conservation, although one can observe that there are contradictions within and across policies, and considerable restrictions around such use. The IUCN's sustainable use policy statement specifies that 'enhancing the sustainability of uses of wild living resources involves an ongoing process of improved management of those resources' (IUCN 2000). The environmental policies that have emerged in India in the past three years send out a very clear message. Irrespective of the declarations in support of sustainable use contained in these, and earlier policies, when and where it really matters, exclusionary conservation trumps sustainable use models of conservation.

Recognition of sustainable use as a conservation tool has remained within the realm of a conceptual device, particularly relevant when ensuring that policy instruments meet accepted international principles. However, when it comes to operative legal provisions that determine how conservation is implemented, it is seen that although sustainable use may have made its way into policy documents it has not made its way into conservation mindsets and practices. In *Wildlife First vs Ministry of Forest and Environment*, conservationists challenged the constitutionality of FRA, which led the Supreme Court in February, 2019 to order the eviction of about a million forest-dependent people whose claims had not been processed under the Act, although this order was subsequently stayed.

Indian conservationists are not alone in their mistrust of sustainable use as a conservation tool. Sustainable use has remained a polarizing policy prescription despite the IUCN Policy Statement on Wild Living Resources (2000) being fairly widely accepted (Hutton and Leader-Williams 2003). The exclusion of people has historically been a part of conservation policies across the world, rooted in the idea that sees humans as distinct from nature (Fairhead and Leach 1996). Even in the face of evidence of better conservation outcomes where ownership and management of forests were devolved to local communities where the ownership and management of forests were devolved to local communities (Blaikie and Springate-Baginski 2007; Lund 2007; Somanathan et al. 2009), the absence of conclusive linkages between conservation and sustainable use is cited as a reason for lack of conviction in it (Hutton and Leader-Williams 2003).

In many circles, however, it is acknowledged that separating people from the governance and use of resources has led to the failure of many a developmental intervention (Saxena 2007). Despite the forceful and well-funded efforts of states and funding agencies to protect forested resources from local populations, they are no match against people whose livelihoods depend on the fodder, fish, and wildlife in forests (Agrawal and Gibson 1999; Sundar 2000; The World Bank 1996). While plans for setting half the world aside for conservation (Wilson 2016) continue to be championed, others warn that there is little space left on the planet for this plan if one continues to be wedded to the ideals of exclusionary conservation (Hutton and Leader-Williams 2003).

As a conceptual tool, the biologically sustainable use of wild resources is largely above reproach. Even the most traditional conservationists would agree that use of resources, wild or otherwise, is inevitable (Hutton and Leader-Williams 2003). At the same time, there is cognition that restricting access of locals to wild resources has come at a cost, and has led to the ineffectualness of many a conservation strategy. Time will tell if sustainable use of wild resources is then the newest attempt to navigate the age-old question of who and what manner of use poses a threat to wild resources – an attempt, to inject a new paradigm into conservation decision-making by actors who control access to such resources.

## Conclusion

As this chapter detailed, through the course of history, different power centres have deployed diverse tactics to dictate access to wild resources in India. The latest in the series is the participatory governance model of JFM and the rights-based approach of FRA. JFM was half-heartedly implemented (Guha 2001), and is acknowledged even by the state as having failed (Narain et al. 2005). Although it saw success in some areas, FRA continues to be scuttled through numerous bureaucratic hurdles and legal challenges (Kutty et al. 2019; Kodiveri 2018).

Newer environmental policies defer to sustainable use in principle, without actually putting it into practice. In much the same mould as the colonial state, while the merits of sustainable use by local populations continues to be deliberated, the state has expanded the legal framework for facilitating the large-scale extraction of wild resources by industry, as evidenced by the draft *Environment Impact Assessment Notification, 2020* (MoEFCC 2020) and the decisions of the Standing Committee of the National Board for Wildlife in April, 2020 (Ravi 2020).

Sustainable use cannot be uncritically adopted as a conservation strategy, given that it is amorphous enough to subsume a range of interventions. Engaging with sustainable use as a conservation model will entail recognition "that a relatively small range of factors have a disproportionate impact on the likelihood that a species or ecosystem will be used sustainably. These vary from the strictly biological, such as the intrinsic rate of population growth, to the socio-economic and political, such as land tenure, access regimes and enforcement" (Hutton and Leader-Williams 2003).

Deployment of this strategy in conservation practice will require that concerns of equity for wild resource users and the conferment of rights are factored in. It would be a disservice to the decades of campaigns and struggle for peoples' participation in resource governance (Sundar 2000) if sustainable use is misappropriated to dilute the gains of peoples' movements across the world. As the trajectory of governance around the sustainable use of the wild unequivocally tells us, resource use is an arena that is intimately entangled with power. Any use of it as a conservation model must therefore necessarily engage with this.

## References

Agrawal, A., & Gibson, C. C. (1999). Enchantment and disenchantment: The role of community in natural resource conservation. *World development*, *27*(4), 629–649. https://doi.org/10.1016/S0305-750X(98)00161-2.

Agrawal, A., & Saberwal, V. (2007). South Asian pastoralism: The environmental question. In M. Rangarajan (Ed.), *Environmental issues in India: A reader*. Delhi: Pearson/Longman. https://doi.org/10.1177/006996671004400218

Blaikie, P., & Springate-Baginski, O. (Eds.). (2007). *Forests, people and power: The political ecology of reform in South Asia*. London: Earthscan. https://doi.org/10.1111/j.1467-7660.2008.00490_3.x

Fairhead, J., & Leach, M. (1996). *Misreading the African landscape: Society and ecology in a forest-savanna mosaic*. Cambridge: Cambridge University Press. doi:10.1017/CBO9781139164023

Guha, R. (2001). The prehistory of community forestry in India. *Environmental History*, *6*(2), 213–238. https://doi.org/10.2307/3985085

Guha, S. (2002). Claims on the commons: Political power and natural resources in pre-colonial India. *The Indian Economic and Social Review History*, *39*(2–3), 181–196. https://doi.org/10.1177/001946460203900204

Guha, S. (2007). A historical study of the control of grass and fodder resources in eighteenth-century Maharashtra. In M. Rangarajan (Ed.), *Environmental issues in India: A reader*. Delhi: Pearson/Longman.

Hutton, J., & Leader-Williams, N. (2003). Sustainable use and incentive-driven conservation: Realigning human and conservation interests. *Oryx*, *37*(2), 215–226. doi:10.1017/S0030605303000395

IUCN. (2000). *The IUCN policy statement on sustainable use of wild living resources adopted at the IUCN World Conservation Congress, Amman*. https://portals.iucn.org/library/efiles/documents/Rep-2000-054.pdf

Kodiveri, A. (2018). Wildlife first, people later? Forest rights and conservation – Towards an experimentalist governance approach. *Journal of Indian Law and Society*, *9*, 39–63.

Kutty, R., Kodiveri, A., Lele, S., & Setty, S. (2019). India's Forest Rights Act, 2006: Stuck in a maze of bureaucratic interpretations. *The Indian Journal of Social Work*, *80*(4), 439–460. DOI:10.32444/IJSW.2019.80.4.439–460

Lal, M. (2007). Iron tools, forest clearance and urbanization in the Gangetic plains. In M. Rangarajan (Ed.), *Environmental issues in India: A reader*. Delhi: Pearson/Longman.

Lund, J. F. (2007). Is small beautiful? Village level taxation of natural resources in Tanzania. *Public Administration and Development*, 27(4), 307–318. https://doi.org/10.1002/pad.467

Ministry of Environment, Forest and Climate Change. (2014). *National biodiversity action plan: Addendum 2014 to NBAP 2008*. www.researchgate.net/publication/280718255_NATIONAL_BIODIVERSITY_ACTION_PLAN_NBAP_ADDENDUM_2014_TO_NBAP_2008

Ministry of Environment, Forest and Climate Change. (2020). *Draft environmental impact assessment notification*.

Mosse, D. (1997). The ideology and politics of community participation: Tank irrigation development in colonial and contemporary Tamil Nadu. In R. D. Grillo & R. L. Stirrat (Eds.), *Discourses of development: Anthropological perspectives* (pp. 255–291). Oxford: Berg.

Narain, S., Singh, S., Panwar, H. S., & Gadgil, M. (2005). *Joining the dots: The report of the tiger task force*. Tiger Task Force, Union Ministry of Environment and Forests (Project Tiger), Government of India. https://ntca.gov.in/assets/uploads/Reports/Joining_the_dot.pdf

Orissa Mining Corporation Ltd vs Ministry of Environment & Forest, Supreme Court of India, Writ Petition (Civil) No. 180 of 2011 (2013).

Press Information Bureau, Government of India. (2019). *Government clears misgivings of amendment in the Indian Forest Act, 1927*. https://pib.gov.in/Pressreleaseshare.aspx?PRID=1591814

Rajan, S. R. (2006). *Modernizing nature: Forestry and imperial eco-development 1800–1950*. Oxford: Oxford University Press. DOI:10.1093/acprof:oso/9780199277964.001.0001

Rangarajan, M. (1996). Environmental histories of South Asia: A review essay. *Environment and History*, 2(2), 129–143. www.jstor.org/stable/20723006

Rangarajan, M. (1998). *Troubled legacy: A brief history of wildlife preservation in India*. New Delhi: Centre for Contemporary Studies, Nehru Memorial Museum and Library, Teen Murti House.

Rangarajan, M. (2007). Introduction. In M. Rangarajan (Ed.), *Environmental issues in India: A reader*. Delhi: Pearson/Longman.

Ravi, P. (Ed.). (2020). *Minutes of 57th meeting of the standing committee of national board for wild life*. Government of India Ministry of Environment, Forest and Climate Change. http://forestsclearance.nic.in/DownloadPdfFile.aspx?FileName=0_0_1111912812141MinutesofNBWLmeeting.pdf&FilePath=../writereaddata/Addinfo/

Saxena, N. C. (2007). Rehabilitating degraded lands. In M. Rangarajan (Ed.), *Environmental issues in India: A reader*. Delhi: Pearson/Longman.

Seventeenth Lok Sabha Secretariat, Committee On Government Assurances, Government of India. (2020). *Second report of the committee on government assurances (2019–2020). Requests for dropping of assurances (not acceded to)*.

Skaria, A. (1992). *A forest polity in western India: The Dangs, 1800s-1920s* [Unpublished doctoral dissertation]. University of Cambridge.

Somanathan, E., Prabhakar, R., & Mehta, B. S. (2009). Decentralization for cost-effective conservation. *Proceedings of the National Academy of Sciences*, *106*(11), 4143–4147. https://doi.org/10.1073/pnas.0810049106

Sundar, N. (2000). Unpacking the 'joint' in joint forest management. *Development and Change*, *31*(1), 255–279. DOI:10.1111/1467-7660.00154

Thapar, R. (2007). Forests and settlements. In M. Rangarajan (Ed.), *Environmental issues in India: A reader*. Delhi: Pearson/Longman.

The World Bank. (1996). *The World Bank participation sourcebook*. https://documents1.worldbank.org/curated/en/289471468741587739/pdf/multi-page.pdf

T.N. Godavarman Thirumulpad vs Union of India, Supreme Court of India, Writ Petition (Civil) No. 202 of 1995 (2005).

Wilson, E. O. (2016). *Half*-earth*: Our planet's fight for life*. New York: Liveright Publishing Corporation.

# 4 Small islands, big lessons

## Critical insights on sustainable fisheries from India's coral atolls

*Naveen Namboothri, Ishaan Khot, and Abel Job Abraham*

### Introduction

Approximately 14 million people in India rely on fisheries and its allied sectors. There are over 3,250 fishing villages dispersed along India's approximately 7,500-km-long coastline with fishing activities spread out along the entire coastline. For centuries, traditional fisher folk have harvested resources from the sea, ensuring a livelihood for themselves while addressing the nutritional requirements of coastal communities. For these traditional communities, fishing is not just a livelihood activity but forms an integral part of their social and cultural identity.

Prior to the mechanization drive of the 1970s and the centralization of fisheries governance, the fisheries sector was mostly decentralized, and marine resource governance was in the form of traditional systems and customary practices that were often regulated at the village level (Kurien 1998). Such traditional common property management regimes often regulated sea tenure, gear restrictions, seasonal closures, and allocation of fishing grounds (Kurien 2000). It provided local communities rights over coastal and marine resources and employed their vast repositories of traditional knowledge, skills, and cooperative fishing techniques (Kurien 1990, Ruddle 1994). These traditional practices have contributed to harvesting resources from the sea sustainably and ensuring the protection of local biodiversity.

From the 1970s onwards, India witnessed a massive government-supported drive to industrialize its fisheries. Mechanization of its fishing fleet, globalization of its markets, and improved post-harvest technology fuelled a massive, unregulated expansion in its fishing efforts (Kurien 1996). Such an unregulated scaling up of its fisheries is reflected in the reduction of its near shore productivity and declines in near shore fishery resources (FAO 2020). A necessary governance change to facilitate this transition was the shift from a common-property based management regime to a centrally managed regime (Kurien 1998, Kurien 2007).

While such transitions helped boost India's fish production and its contribution to the country's Gross Domestic Product (GDP), it shifted power from local communities to the state. Fishers were reduced to being passive

DOI: 10.4324/9781003343493-5

participants or silent producers with no regulatory powers or roles in decision making. The stripping of power away from fisher communities left them with no jurisdiction over their fishing grounds and the decline of traditional governance systems. Contemporary state-owned fisheries management regimes failed to provide spaces for local communities or their local ecological knowledge, or to contribute to fisheries management and governance (Kurien 2007).

## Dichotomies and incompatibilities

This deep-rooted epistemological dichotomy between the contemporary science-based, production-oriented fisheries management regimes and the traditional, local ecological knowledge (LEK)-based, decentralized governance regimes has often led to systemic incompatibility between the two (Pomeroy 1995, Jentoft et al. 1998). While the former has evidently failed in stemming the rampant decline in fisheries resources globally (McGoodwin 1991) and in some cases exacerbated it (Marchak et al. 1987, Hannesson 1996), the latter has been significantly eroded under the onslaught of globalization and centralization of the fisheries sector (Kurien 2007).

It is increasingly evident that fisheries management models of temperate developed nations that rely on arriving at the maximum sustainable yield (MSY) of a few key fish species and centralizing administrative authority are incompatible with the fisheries of the tropical developing nations, which are characterized by highly dispersed, diverse, and complex fisheries (Pomeroy 1995). It is becoming evident that states cannot single-handedly regulate and manage fisheries without decentralizing fisheries governance and actively engaging communities.

Studies focusing on fisheries governance in many developing countries have identified that resources can be better managed when local community members actively participate in management decisions and access and decision-making rights are distributed more equitably (Pomeroy 1995, McCay and Jentoft 1996, Lobe and Berkes 2004). Along with the promotion of participatory mechanisms of fisheries management, there is renewed interest in learning from traditional systems of governance where biodiversity conservation and community needs go hand in hand (Kuperan and Abdullah 1994, Kurien 1996). Co-management – a bottom-up, decentralized approach – where resource users are granted both rights and responsibilities through delegation of management authority, is increasingly being promoted as a method to address the dichotomy (Jentoft et al. 1998).

With a case study from the Lakshadweep Islands, this chapter aims to demonstrate how traditional governance systems rooted deeply in LEK can ensure that sustainable livelihoods and ecosystems can co-exist, and also how the incompatibility between the traditional and the contemporary forms of fisheries management could lead to unwanted transitions that push fisheries down paths of unsustainability. We also summarize some of Dakshin Foundation's attempts in Lakshadweep to bridge the gap between the contemporary and the traditional by a) democratizing the practice of science and encouraging the local fisher community to be the creators and owners of scientific information regarding their fisheries and b) building spaces that facilitate a co-management regime, thereby empowering local fisher communities to actively participate in fisheries management decisions.

## Sustainable use in India's coral islands

The Lakshadweep Islands are India's only coral atolls and are home to about 65,000 people who directly or indirectly depend on the reef and the ocean around them. The main fishery practised here – the 'live-bait pole-and-line tuna fishery' is a unique, best-practice fishery that targets the resilient skipjack tuna in a low-impact, selective manner that diverts fishing pressure off the sensitive coral reefs that constitute these atolls (Arthur 2013, Alonso et al. 2015, Jaini et al. 2018). Additionally, being a labour-intensive technique, it is one of the major sources of livelihoods for the local community in Lakshadweep. It may thus be one of the last remaining examples of a sustainable commercial fishery in India (Figure 4.1).

A critical limiting factor for the pole-and-line fishing operations is the availability of adequate amounts of baitfish (Kumar et al. 2017). Baitfish is a small fish that is found in the lagoons and reefs around the islands that are used as bait for catching tuna. Fishermen catch the baitfish and keep them

*Figure 4.1* Pole-and-line fishing for skipjack tuna as practised in Lakshadweep
*Illustration: Ananya Singh*

alive in on-board holding tanks before heading out to catch tuna in deeper waters, giving the technique its name – 'live-bait' pole-and-line fishing. Our work in Lakshadweep has focused primarily on addressing issues around baitfish availability and management.

Pole-and-line fishery has its origins in the Maldives. From the Maldives, this fishing technique (or fishery) came to Minicoy, the southernmost island of the Lakshadweep archipelago a few centuries ago (Jaini and Hisham 2013), although the exact period and mode of arrival is not clear. Over the centuries, the native population of Minicoy evolved elaborate systems of customary practices and institutions to govern their commons, both on the land and in the sea. Many of these practices demonstrate a strong sense of stewardship with regard to their common-pool resources, a nuanced understanding of the ecological fragility of the islands and an understanding of the biology and behaviour of the species that are harvested.

Until the 1960s, the rest of the nine inhabited islands of Lakshadweep archipelago practised only artisanal fisheries targeting reefs and other off-shore resources (Union Territory of Lakshadweep, Department of Fisheries 1990). In the 1960s, the Fisheries Department, along with the introduction of mechanization, transferred the practice of pole-and-line tuna fishing from Minicoy to the rest of the islands (Hoon 2003). Since then, this has been the major fishery practised in the islands contributing to over 90% of tuna landings in Lakshadweep (Vinay et al. 2017).

However, fisheries managers trained in contemporary methods of resource management and encouraged by the promise of tapping into the extensive tuna resources surrounding the Lakshadweep islands, failed to recognize the immense wealth of traditional knowledge and customary practices that were in place on Minicoy that supported a sustainable resource use regime. Our interviews with the senior fishermen from Lakshadweep indicate that while the technique of pole-and-line tuna fishing was introduced from Minicoy to the other islands, the associated management practices were ignored. In the following section, we describe some of the traditional fisheries management practices in place on the island of Minicoy, an understanding of which is crucial to demonstrate the inherent incompatibilities between the contemporary and traditional forms of fisheries governance.

## Traditional systems of resource management on Minicoy Island

The geographical isolation of Minicoy Island from the rest of the archipelago meant that the traditional knowledge systems and customary practices on these islands evolved in relative isolation. While such practices and knowledge systems addressed multiple facets of island governance including food and water security, land tenure, dispute resolution, social events, profit sharing, and so on, in this section, we focus on a few examples of customary resource management practices and the associated TEK related to the island's fisheries, particularly baitfish resource management.

### *Customary institutions*

The fisheries Jamaat is a customary institution of Minicoy which decides on local rules and management practices related to fishery resources (Figure 4.2). Currently, it is a formally registered fisheries union known as the Maliku Masverin Jamaat where the decisions are taken democratically through a consultation process between the owners and Kelus (boat captains) of all pole-and-line tuna boats of Minicoy.

Another customary institution is Minicoy's 'village system' wherein the populace of the island is divided into 11 villages (known as Avah), each headed by two Moopans (male heads) and two Moopathis (female heads). The Avah is an economically active unit responsible for the governance of common-pool resources on the island such as common lands, ponds, fishing boats, coconut trees, etc. that come under its jurisdiction.

### *Baitfish resource management systems on Minicoy Island*

Since baitfish is one of the critical resources for the pole-and-line fishery and is prone to over-exploitation and rapid population declines, its management has received considerable attention in Minicoy's customary practices. This includes spatial and temporal regulations that demonstrate a nuanced understanding of the biology and ecology of the baitfish species.

*Figure 4.2* Meeting with representatives of the Minicoy fisheries Jamaat, May 2018
*Photo: Mahaboob Khan*

We summarize below some of the key findings from our studies which also corroborate previous studies on the subject.

- Seasonal bans on baitfish collection: There is a seasonal ban on catching a particular variety of baitfish, locally known as Bodhi (Apogons) during its breeding season from May to September. In the past, a strict monsoon ban on the harvest of all species of baitfish used to be enforced, providing their populations time to recover. In order to ensure the enforcement of this practice, the Jamaat would essentially restrict large-scale tuna fishing during the monsoon in Lakshadweep, allowing only small boats to operate.
- Rules for baitfish harvesting and use: Baitfish varieties, locally known as Hondeli and Rahi (species of sprats) are caught only from the shallow sandy parts of the southern side of the Minicoy lagoon where coral presence is sparse. Such practices are believed to minimize damage to corals while hauling the nets and ensure minimum habitat degradation. Similarly, a species of sardine, locally known as Kumbala, is not used as bait in the open ocean as it is believed that the scales of Kumbala can damage the tuna's stomach lining.
- Quantification and conservation of baitfish: While the quantification of tuna resources that are harvested is relatively easy, the quantification of baitfish harvest has been a major challenge. However, Minicoy has developed species-specific units for estimating the extent of baitfish caught. These units, which are usually based on the fishing gear or the size of the baitfish box, enable an approximate measurement of the baitfish catch volumes/weights. Also, locally designed in-water baitfish tanks known as Labari are often used for storing leftover baitfish after each fishing trip, thereby minimizing wastage (Figure 4.3).
- Resource allocation: Magaos are large coral boulders, located within the lagoon, that act as Bodhi (apogons) banks for individual boats. Before the onset of the fishing season, individual boats are permitted to select a particular Magao of their choice at a Jamaat meeting attended by the owners and Kelus of all boats. While boats are free to harvest bait from unassigned coral boulders, a Magao designated to a certain boat is out of bounds for other boats.

In addition to the specific rules regarding baitfish utilization and management, there exists a great wealth of knowledge and customary practices vis-à-vis other aspects of fisheries such as post-harvest processing of tuna, and navigation practices based on a detailed understanding of current patterns and astronomical phenomena (Hoon 2003).

## Contemporary resource use and governance

Current resource use and governance scenario of fisheries in Lakshadweep reflect the dichotomy mentioned earlier and its consequences. Governance is primarily top-down like in India, offering very little space or opportunities

*Figure 1.3* A modern variant of Labari used for in-water storage of leftover baitfish
*Photo: Shweta Nair*

for communities to officially engage with the system. The Fisheries Department through its units on each island is responsible for managing the day-to-day affairs of fisheries such as supplying diesel to fishers, licensing and issuing fishing permits and implementing welfare schemes and subsidies of the government. The dominant institutional perspective of developing fisheries by focusing singularly on increasing capacity to augment fisheries production and maximize revenue from the oceans is passed down from the national level and is echoed at local levels of administration as well.

At the community level, institutions such as Minicoy's fisher Jamaat still have some decision-making powers related to fisheries management. While there is general acknowledgement from both fishers and the government about the existence and influence of the Jamaat, the lack of formal recognition gives rise to legitimacy concerns and poses limitations to the degree to which the Jamaat can engage in matters of fisheries management. While traditional institutions like the Jamaat do not exist in islands other than Minicoy, there have been some sporadic attempts from fishers to organize and unionize. Driven by enterprising individuals from fishing communities, different local fisher unions have had varying degrees of success over the years; however, they have eventually broken down due to political differences or a lack of strong leadership.

The past four to five years in Lakshadweep have seen the fishery go through significant transitions driven by a variety of factors. In addition to pole-and-line fishing for skipjack tuna, there has been a surge in handline fishing for yellowfin tuna since 2017–18 after the Fisheries Department started allowing collection boats from the mainland to collect fresh fish from Lakshadweep. While this is a boon for Lakshadweep fishers as it provides them instant cash, fixed rates, and saves them the labour of making masmin (Lakshadweep's specialized dried tuna product), it is also indicative of how volatile these systems are and how a single development can drive changes in resource use patterns.

Maldivian influence on fisheries is also increasingly being seen, evident through videos of Maldivian fishing techniques and fishing boats that are commonly passed around among fishers on their smartphones. These are often considered to be the standards to aspire for, especially by the youth. Encouraged by government subsidies that aim at increasing production, there has been a strong upward trend in constructing bigger fishing vessels capable of multi-day fishing that have led to an increase in overall fishing capacity. The differences in boat size have translated to a differential power dynamic within the community to harvest a common pool of resources. In what was earlier a fairly homogeneous fishery in terms of scale of operations, a small-scale versus large-scale distinction is now increasingly evident.

One of the most significant consequences of these factors has been the gradual precipitation of a baitfish crisis in the islands over the years. Practices such as light fishing (fishing using LED lights or torches to catch certain species of baitfish in the night) or using small-meshed nets to catch baitfish, inspired partly by Maldivian fishing techniques, and fuelled by competition due to increased fishing capacity, are now rampant. Such practices target baitfish stocks before they can spawn and are therefore potentially unsustainable. While fishers are aware of the deleterious impacts of such practices, they are forced to participate in them due to a competitive situation and lack of regulations. While the problem is common across Lakshadweep, it manifests in varying degrees on different islands.

This has led to a situation where a traditionally sustainable fishery is at the risk of going down unsustainable pathways driven by specific practices and management approaches. In a common pool resource crisis, traditional management systems with their inherent sustainability concerns can act as a strong counterweight and also enable resilience. While problems of overcapacity and resultant competition exist in Minicoy as well, the presence of strong customary management systems in Minicoy has enabled it to cope with issues of baitfish decline better than other islands.

However, the absence of such practices and associated knowledge in the other islands has rendered them more vulnerable to the impacts of external shocks and sudden changes. It is important to note here that while Minicoy may fare better than other islands in terms of coping with changes and unsustainability, the lack of formal recognition for its traditional knowledge systems and the erosion over time of these systems due to disinterest at the local level have created their own set of challenges. The prioritization of state systems of governance over the customary systems and their inherent incompatibility has made enforcement of certain practices such as penalizing the violators of customary rules difficult.

## Strengthening sustainable use: Dakshin's community-based efforts in Lakshadweep

Dakshin Foundation has been working in Lakshadweep since November 2012 with the aim of studying and preserving the sustainable use systems in the islands, particularly the pole-and-line tuna fishery. Through interdisciplinary action research on the social, ecological, and economic aspects of island fisheries and contextualized interventions with the community, our work seeks to empower the local fishing community to take knowledge-based actions for sustainability.

With the objective of bridging the gulf between traditional and modern systems of knowledge, we decided to try and leverage the nascent potential of communities in understanding local fisheries better and launched a community-based fisheries monitoring (CBFM) programme in January 2014 in the islands of Agatti, Kadmat, Kavaratti, and in Minicoy in 2015 (Figure 4.4).

*Figure 4.4* Launch of community-based fisheries monitoring programme in Agatti

*Photo: Keerthi Chinnappa*

The programme involves active fishers regularly monitoring pole-and-line fishery dynamics for the long term.

Fishing communities' interface with the ocean on a daily basis and traditionally have a plethora of observations and knowledge about the ocean and fisheries. But due to eroding traditional knowledge systems and lack of systematic documentation, these observations remain unacknowledged (Sridhar and Namboothri 2012). CBFM is an attempt to mainstream these observations and knowledge in a modern science framework (Shanker and Oommen 2018). As a monitoring approach, CBFM offers several advantages – it enables fine-scale data collection at a much larger spatial and temporal scale than independent researcher/enumerator-led monitoring. It is also a cost-effective tool for large-scale data generation on fisheries. Most importantly, it enables fishers to see patterns in their fishery over time, enabling them to make local-level decisions about fisheries management. This participatory approach seeks to decentralize the process of knowledge generation and reduce fishers' reliance on external agencies for information about their own fisheries (Figure 4.5).

Keeping with the participatory spirit of this approach, CBFM logbooks were developed in consultation with active fishers. To date, a total of 50 boats have participated in this programme, collectively contributing over 4,000 fishing records to a community-generated dataset on island fisheries and demonstrating the potential of local communities to engage constructively in natural resource monitoring activities.

Envisaged as a two-way knowledge sharing platform, the data collected through the programme is returned to fishers via simplified reports containing boat-level and island-level fishery metrics. Interactions with fishers who actively participate in the CBFM programme reveal that keeping detailed records of day-to-day fishery dynamics helps them track parameters like diesel consumption and baitfish utilization over time and manage fishing operations efficiently.

*Figure 4.5* Infographic depicting CBFM

*Illustration: Prabha Mallya*

## From monitoring to management

In addition to data, the CBFM programme has helped create a strong network and build trust with the local fishing community which is now enabling broader dialogues around resource management, especially critical pressing issues on the ground such as the baitfish crisis and light fishing. Efforts over the past two years have been geared towards creating a fisheries co-management framework. Co-management can serve as an effective platform for various stakeholders (fishers, government agencies, scientists, NGOs, traders, etc.) to engage with each other on various issues related to local fisheries and serve as a tool for democratic decision making (Pomeroy 1995, Sen and Nielsen 1996, Jentoft et al. 1998).

After a year of scoping and feasibility surveys with key informants and active fishers, Dakshin Foundation launched a fisheries co-management initiative in May 2019. Through three stakeholder consultation meetings held in Kavaratti, Agatti, and Minicoy, the basic concepts of co-management were introduced to fishers and representatives of the fisheries department on each island. To demonstrate co-management in action, light fishing for baitfish was taken up for discussion. After much deliberation and mediation, fishers on all the three islands resolved to give up the practice of light fishing. Other ecologically destructive practices such as dumping of fish waste in island lagoons were also discussed and consensually curtailed by fishers with the support of the Fisheries Department (Figure 4.6). In Minicoy, fishers also discussed and agreed to phase out the use of small-meshed nets for catching baitfish.

*Figure 4.6* Launch of Dakshin's fisheries co-management initiative in Kavaratti, May 2019

While this was an optimistic and promising start, there is a lot of ground to cover for these interventions to be sustainable in the long run. Within a few months of the first round of meetings, we received reports from the ground that many fishers had recommenced light fishing again in some places, whereas in others, the collective decisions were being respected and followed. A deeper understanding of community dynamics and local politics as well as sustained engagement and follow-ups with all actors involved will be required to ensure the intended outcomes from such community-based initiatives.

## Conclusion

In a world that is increasingly dominated by neo-liberal developmental paradigms and resultant environmental degradation, examples of balancing human needs and biodiversity have become few and far between. It is essential that such examples, wherever they still exist, are well-documented and supported. The Lakshadweep Islands with their unique pole-and-line fishery present one such example. They also offer an opportunity to preserve and strengthen a traditionally sustainable use practice and prevent it from transitioning down unsustainable trajectories. The islands' history of maritime dependence, associated knowledge practices, relative isolation, and strong social cohesion among the local communities have buffered its fisheries from external influences this far.

Our experience working in Lakshadweep over the past eight years, however, has shown that factors such as increased exposure, external market forces, and context-independent fisheries; these development paradigms are creating disparities within the community and a departure from a sustainable use scenario. The baitfish crisis example clearly demonstrates the inability of the contemporary management regime to acknowledge and integrate a nuanced and locally successful traditional management regime. Such deep-rooted systemic incompatibilities between contemporary and traditional fisheries management regimes have now placed Lakshadweep on the threshold of an unsustainable trajectory.

Given the inevitability of such transitions, not just in the fisheries sector but also in Lakshadweep's overall development, Dakshin Foundation's attempts to build hybrid models of governance and bottom-up participatory processes are more than just an attempt at sustainably developing its fisheries. We hope the lessons learned from these attempts can be applied to all aspects of Lakshadweep's development. The Lakshadweep Islands, given their social and ecological homogeneity and high levels of community literacy, are ideal to test out alternative approaches to governance and development. They have the potential to be developed as a bright spot in the governance of small-scale fisheries in India.

## Acknowledgements

Our work in the Lakshadweep Islands is possible thanks to the support of the Lakshadweep Administration, Department of Science and Technology, Department of Fisheries and the local fishing community. We are extremely grateful to Tata Trusts, Blue Ventures, and Rohini Nilekani Philanthropies for supporting this work over the years.

## References

Alonso, D., Pinyol-Gallemí, A., Alcoverro, T., & Arthur, R. (2015). Fish community reassembly after a coral mass mortality: Higher trophic groups are subject to increased rates of extinction. *Ecology Letters*, *18*(5), 451–461. DOI:10.1111/ele.12426

Arthur, R. (2013). Happenstance and the accidental resilience of the Lakshadweep Reefs. *Current Conservation*, 7(2), 30–35. www.currentconservation.org/happenstance-and-the-accidental-resilience-of-the-lakshadweep-reefs-2/

FAO. (2020). *The state of world fisheries and aquaculture 2020: Sustainability in action*. Rome: FAO. https://doi.org/10.4060/ca9229en

Hannesson, R. (1996). *Fisheries mismanagement: The case of the Atlantic Cod*. New York: Wiley-Blackwell.

Hoon, V. (2003). *Socio-economic dimensions and action plan for conservation of coastal resources based on an understanding of anthropogenic threats*. Chennai, India: Centre for Action Research on Environment Science and Society.

Jaini, M., Advani, S., Shanker, K., Oommen, M. A., & Namboothri, N. (2018). History, culture, infrastructure and export markets shape fisheries and reef accessibility in India's contrasting oceanic islands. *Environmental Conservation*, *45*(1), 41–48.

Jaini, M., & Hisham, J. (2013, December 10–13). *Sustainable pole and line tuna fisheries in the Indian Ocean: Does Lakshadweep hold up to Maldives' MSC standards?* [Paper presentation]. International conference on small-scale fisheries governance: Development for wellbeing and sustainability, Hyderabad, India. www.dakshin.org/wp-content/uploads/2017/10/MSC_JainiHisham2013.pdf

Jentoft, S., McCay, B. J., & Wilson, D.C. (1998). Social theory and fisheries comanagement. *Marine Policy*, 22(4–5), 423–436.

Kumar, K. V., Pravin, P., Meenakumari, B., & Boopendranath, M. R. (2017). Fishing craft and gears of Lakshadweep islands – A review. *South Indian Journal of Biological Sciences*, *3*(1), 1–6.

Kuperan, K., & Abdullah, N. M. R. (1994). Small-scale coastal fisheries and co-management. *Marine Policy*, *18*(4), 306–313. https://doi.org/10.1016/0308-597X(94)90045-0

Kurien, J. (1990). Knowledge systems and fishery resource decline: A historical perspective. In W. Lenz & M. Deacon (Eds.), *Ocean sciences: Their history and relations to man. Proceedings of the 4th International Congress on the history of oceanography* (pp. 476–480). Hamburg: Deutsches Hydrographisches Institut.

Kurien, J. (1996). *Towards a new agenda for sustainable small-scale fisheries development*. South Indian Federation of Fishermen Societies. https://dlc.dlib.indiana.edu/dlc/bitstream/handle/10535/5118/Towards%20a%20new%20agenda%20for%20sustainable%20small%20scale%20fisheries.pdf?sequence=1

Kurien, J. (1998). Small-scale fisheries in the context of globalization. *Center for Development Studies*. www.researchgate.net/publication/5127025_Small-scale_fisheries_in_the_context_of_globalisation

Kurien, J. (2000). Community property rights: Re-establishing them for a secure future for small-scale fisheries, FAO Fisheries Technical Paper 404/1. In R. Shotton (Ed.), *Use of property rights in fisheries management* (pp. 288–294). Rome: FAO.

Kurien, J. (2007). The blessing of the commons: Small scale fisheries, community property rights and coastal natural assets. In J. K. Boyce, S. Narain, & E. A. Stanton (Eds.), *Reclaiming nature: Environmental justice and ecological restoration* (pp. 23–54). London and New York: Anthem Press.

Lobe, K, & Berkes, F. (2004). The padu system of community-based fisheries management: Change and local institutional innovation in south India. *Marine Policy, 28*(3), 271–281.

Marchak, P., Guppy, N., & McMullan, J. (Eds.). (1987). *Uncommon property: The fishing and fish-processing industries in British Columbia*. Methuen Publications. https://doi.org/10.14288/bcs.v0i88.185847

McCay, B. J., & Jentoft, S. (1996). From the bottom up: Participatory issues in fisheries management. *Society and Natural Resources*, *9*(3), 237–250. https://doi.org/10.1080/08941929609380969

McGoodwin, J. R. (1991). *Crisis in the world's fisheries: People, problems, and policies*. Stanford, CA: Stanford University Press.

Pomeroy, R. S. (1995). Community-based and co-management institutions for sustainable coastal fisheries management in Southeast Asia. *Ocean and Coastal Management*, 27(3), 143–162. https://doi.org/10.1016/0964-5691(95)00042-9

Ruddle, K. (1994). *A guide to the literature on traditional community-based fishery management in the Asia-Pacific tropics*. FAO. www.fao.org/documents/card/en/c/c215cabf-b01e-5438-84ca-ee508acf0c14

Sen, S., & Nielsen, R. (1996). Fisheries co-management: A comparative analysis. *Marine Policy*, *20*(5), 405–418.

Shanker, K., & Oommen, M. A. (2018, October 25). *Engaging communities in resource monitoring: The political ecology of science as the language of power*. Radical Ecological Democracy. https://radicalecologicaldemocracy.org/engaging-communities-in-resource-monitoring-the-political-ecology-of-science-as-the-language-of-power/.

Sridhar, A., & Namboothri, N. (2012). *Monitoring with logic and illogic: A case for democratising observation in fisheries*. Dakshin Foundation and Foundation for Ecological Security. www.dakshin.org/wp-content/uploads/2017/06/Monitoring_LogicIllogic.pdf

Union Territory of Lakshadweep, Department of Fisheries. (1990). *Thirty years of fisheries development in Lakshadweep*. Kavaratti, Lakshadweep: Department of Fisheries.

Vinay, A., Ramasubramanian, V., Krishnan, M., Kumar, N. R., & Ayoob, A.E. (2017). Economic analysis of tuna pole and line fisheries in Lakshadweep. *Indian Journal of Geo Marine Sciences*, *46*(5), 947–957.

# Part II
# Enterprises

# 5 Ensuring sustainable harvests through market-based tools and community-based organizations

## A practitioner's perspective

*Snehlata Nath*

## Introduction

Harvest of forest produce has historically been prevalent among indigenous communities. Most of their cultural traditions, subsistence needs, and barter exchanges were related to these forest products. It was in the late 1980s that this group of products from the forest, which were previously called 'minor forest produce', came to be defined as 'non-timber forest produce' (NTFP). Well-managed NTFP systems became flagships of a sustainable development model for poor marginalized communities who lived in resource-rich areas (Shackleton et al. 2011). Since sustainability incorporates a balanced and holistic view of the planet, profit, and people, these three aspects (social, environmental, and economic) also formed a part of NTFP interventions (Ticktin and Shackleton 2011). The representation below shows the various elements involved, their overlaps and interactions (Figure 5.1). A cycle of this sort of development brings long-term benefits to all the three aspects in an ideal world.

Over time, with commercial interests in this sector increasing, NTFPs began to have more economic value. The threat of unsustainable harvest increased with the growth of industries using NTFPs, foremost among them being the herbal medicine and cosmetics industry. The scale and magnitude of NTFP collection can be seen from some estimates put forward by researchers. An estimated one in six persons globally depends on forests, particularly for supplementary food, including about 60 million indigenous people who are almost wholly forest dependent (Mansourian et al. 2015). It is estimated that 275 million rural poor, that is about 27 percent of India's total population, depends on NTFPs for at least part of their subsistence and cash livelihoods (Malhotra and Bhattacharya 2010; Bhattacharya and Hayat 2009).

This dependency is intense for half of India's 89 million tribal people, who live in forest fringe areas. Estimates of the revenue contributions of NTFPs in India vary considerably. Some estimate that NTFPs contribute US$208 million to the Indian economy, while another calculation places the

DOI: 10.4324/9781003343493-7

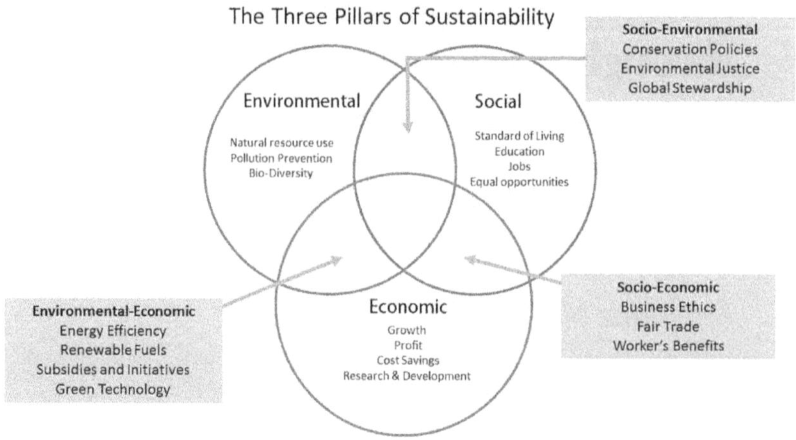


*Figure 5.1* The three pillars of sustainability

Source: Students' Union University for the Creative Arts

revenues from NTFP at US$645 million (Lele et al. 2010). Yet another estimate offered by Poffenberger (1994), finds the total annual contribution of NTFPs from the central Indian tribal belt to be in excess of $500 million. All of these estimates underscore the economic significance of collection and processing of NTFPs to forest-dependent communities.

Given this extent of collection and associated pressures on forests for these resources, the future for 'sustainable use' seems unachievable. This chapter seeks to explore, through various examples of organized NTFP resource management across the country, how different variables and stakeholders influence sustainable use of NTFPs. In doing so, it aims to showcase scenarios and possibilities for the sustainable use of NTFPs.

## Underlying factors of sustainable NTFP management

### *Daily wage and forest use*

For many tribal families, NTFP collection is important and contributes significantly to income (Figure 5.2). It is now well established that about 10% to 40% of family income of harvesters comes from NTFPs (Pain 2009). In the central Indian states, Odisha, and Andhra Pradesh, seasonal collections of some NTFPs such as tendu leaves, mahua flowers and seed, and sal leaf are a regular part of the people's livelihoods. As these products are seemingly found in abundance, their sustainability does not rank high as a concern.

*Figure 5.2* NTFP processing in progress

There is a lack of authenticated information on the value of NTFPs that are sold by indigenous people in *haats* or local markets. These markets are unregulated, informal, and volatile and offer low prices for the harvesters. This low-value, high-volume market also means that the NTFP gatherer makes daily estimates of the income she/he will make for the NTFPs collected and aims to acquire the necessary quantity, with little or no regard for sustainability.

For instance, the low rates paid for wild tender mango meant that harvesters lopped branches and quickly collected sacks of the unripe fruit to sell to the agent in the local market. The lopping of the branches has implications for the health of the tree, and the removal of young fruit in large quantities will certainly impact the population dynamics in the long term. For tribal communities, the wage they earn today has become more important than the long-term sustainability of the NTFP they harvest. This is most often the case with families in poverty and a lack of tenurial security over the resource.

### *Traditional boundaries, tenure, and stewardship*

The role of tenure in commercial NTFP collection is an important factor. In most cases, commercialization and industrial-scale use of NTFPs has led to high demand and has forced communities to harvest extensively. This form of demand places no preconditions on harvesting practices or the source of resources. It leads to local resource competition where the harvester who can take the produce first benefits the most. In certain cases, ancestral domains and boundaries determine the limits of collection for a village, but this is becoming increasingly difficult to follow when cash returns are high. Where NTFP is privately owned, as seen in the case of Mahua trees on farmlands, the ownership is clear and sustainable practices are followed. In forests, which are common property resources, collection is open to all, leading to depletion, destructive harvesting methods, and degradation of the overall habitat. This creates a problem for resource stewardship and overall sustainable development.

### *Plant parts harvested and population*

The sustainability of commercially harvested NTFPs also depends on the part of the plant being harvested and on the life cycle of the harvested species (Ticktin 2004). Whole plants like *Kurunthotti* (*Sida cordifolia)* in the Western Ghats are widely collected for the ayurvedic industry. The scale of collection is an important factor too. A specific study on the annual requirement of *Kurunthotti* among 750 registered manufacturers was shown to be more than 200 tons (Harilal 2009). If unsustainably harvested before flowering and seed dispersal, the shrub would soon disappear from the forest. Similarly, other sensitive plant parts like resins or barks must be harvested sustainably to ensure a healthy population of the NTFP. Research institutions have determined sustainable methods of harvesting of all NTFPs, but their practical application and monitoring have been difficult.

### *Transparency and markets*

The nature of the market and its level of formality also determine sustainable harvesting of NTFPS. In the informal market, traders have a strong hold and operate through a system of agents (Shackleton et al. 2015). They exert control over the information circulating within the markets, which in many cases, facilitates illegal trade. The harvests of Eendh (*Cycas circinalis*) leaves, cones, seeds, and pith from the forests of Western Ghats are a good example of this. Chains of traders, wholesalers, and retailers dealing with the Eendh plant parts exist in most parts of India (Krishnamurthy et al. 2013). They can be seen in the NTFP markets at Amritsar and Khari Bawli in Delhi. Raipur in central India and Dindigul, Tuticorin in Tamil Nadu are also hubs of this informal trade.

This system of marketing has been hard to change as there are old relationships between traders and communities. People depend on the traders for emergencies, loans, and other support which keeps them indebted to the trader. This form of market places demands only on the volume – quality is managed through creating grades based on the quality and texture of the product. Char (*Buchanania lanzan*) in Central India and Siyadi/Adda (*Bahunia vahlii*) leaf in Eastern Ghats are unsustainably harvested only because the demand is high. In the process of collection, immature fruits and leaves are also harvested only to be thrown away. The profit margins are small for the traders, and therefore, volume becomes an important factor.

With the intentions of breaking this exploitative trend, Mutually Aided Co-operative Societies (MACS), Large scale Multi-Purpose Societies (LAMPS), and Tribal Cooperative Societies were set up by state governments to benefit the gatherer and provide fair prices (Setty et al. 2008). Here again the focus was on wages, with prices offered to harvesters with scant regard for ecological sustainability. Several NGOs across India have made efforts to form peoples' marketing initiatives of Producer Companies or Self Help Group (SHG) Federations. While several were successful in removing the chain of traders and getting better returns to the harvesters directly, once again ecological sustainability remained secondary (Shanker et al. 2005).

### *NTFP governance and rights*

Within the past decade, efforts have been made by government and related agencies to formalize institutions which will manage and govern NTFPs. Despite declaring a minimum support price (MSP) and introducing schemes like the 'Van Dhan Yojana', people have not been granted ownership or responsibility for NTFPs. The *Scheduled Tribes and Other Traditional Forest Dwellers (Recognition of Forest Rights) Act, 2006* (FRA) has ensured that NTFPs are a peoples' resource, but years of alienation from forests due to restriction of access is making it difficult for communities to take over NTFP or forest governance.

In some areas, largely due to efforts of NGOs, newly formed Forest Rights Committees have been able to make headway in NTFP management. These successes were fuelled by strengthening local governance of Gram Sabhas, especially in areas to which Panchayats (Extension to Scheduled Areas) Act, 1996 (PESA) applies. Without the granting of rights over resources by the government, it is difficult to ensure sustainable use. Experiences from across the world define local governance as a key ingredient to conservation of forests, alongside the balanced extraction of forest resources.

### *Need for multidimensional data*

Information, data, and studies on NTFP are increasingly being developed in the last decade across the world. However, lack of protocols set in data collection, paucity of ecological studies and a lack of understanding of indigenous cultures have made the issues related to NTFPs complex to comprehend. Therefore, it is hard to analyse data to see if commercially harvested NTFPs follow principles of sustainability. Often the only data that exists comes from marketing agencies, and even though it is not comprehensive, it is the only source of information available to use. Tendu Patta in Madhya Pradesh is one of the main commercial NTFPs collected. Even though abundantly available, the data from government sources within Madhya Pradesh and Chhattisgarh (see Figure 5.3, Figure 5.4, and Figure 5.5) show decreasing quantity trends for Tendu leaf, Terminalia chebula fruit, and Gums.

With limited data, there is not enough evidence or research available to determine the reason for this declining trend. There could be many factors at play – importantly that the consumption of *beedis* has reduced, leading to less demand for Tendu leaves. Local NTFP collectors report that the use of fire during harvest has impacted saplings, increased mortality, and affected tree populations. Furthermore, there has been outmigration of harvesters to cities in search of higher paid jobs, which may also be an important contributing factor to these observed patterns of declining Tendu leaf harvest. Similarly, for Gums or fruit of Chebula collected, a declining trend is evident, although difficult to comprehend, since the demand for these NTFPs has not reduced and the ayurvedic industry which depends on them is predicted to grow very rapidly.

Therefore, a need exists for multidimensional information on NTFP population dynamics, demand, marketing mechanism, tenure, governance, institutional mechanisms, and harvesting practices in order to obtain the

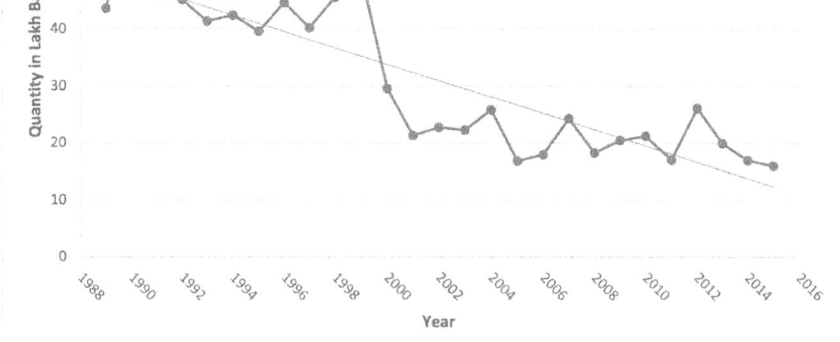


*Figure 5.3* Tendu leaf collection in Madhya Pradesh

Source: Madhya Pradesh State Minor Forest Produce (Trading & Development) Cooperation Federation

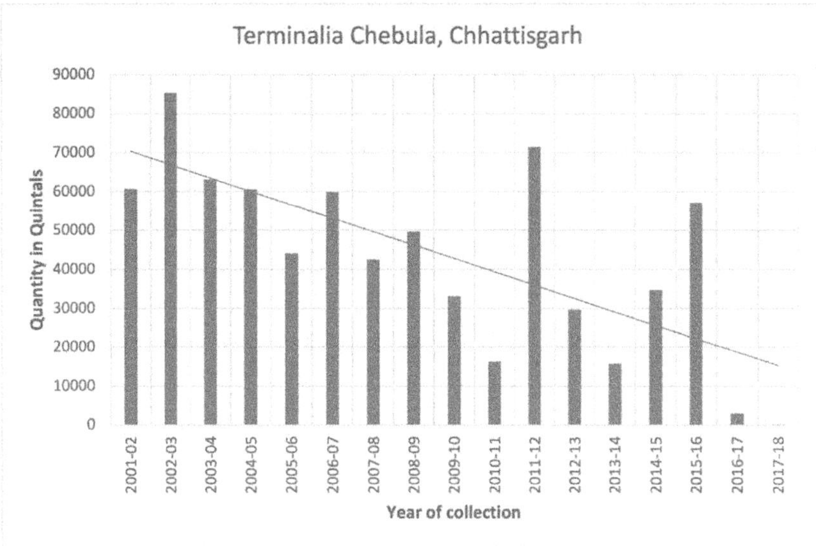


*Figure 5.4* Terminalia chebula fruit collection in Chhattisgarh

Source: Chhattisgarh State Minor Forest Produce (Trading & Development) Cooperation Federation

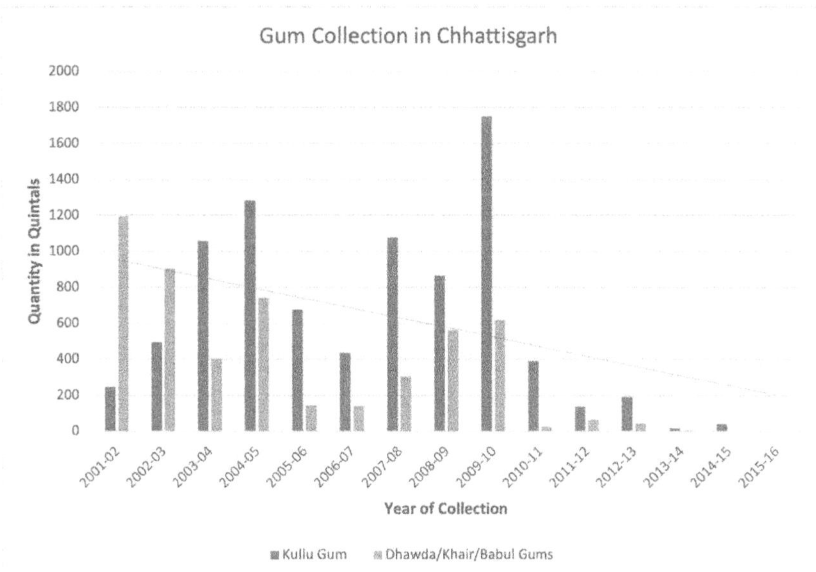


*Figure 5.5* Gum collection in Chhattisgarh

Source: Chhattisgarh State Minor Forest Produce (Trading & Development) Cooperation Federation

entire picture of sustainable use. Such studies would also need to factor in the variability of the resource regime since ecology, economy, and culture play important roles in determining sustainability. Through the examples of two regions and the interventions made for NTFP management, these and other factors that contribute towards sustainable management are discussed further.

## Interventions for NTFP management

### *Eastern Ghats, India – interventions by Kovel Foundation for sustainable NTFP management*

Sustainability of medicinal plants collection from the wild is a matter of concern and in many cases, indiscriminate overharvesting has driven plants from abundance to near endangered status. Some of the well-known cases include Gum karaya (*Sterculia urens*) in Andhra Pradesh, Safed Musli (*Chlorophytum borilivianum*), Arjuna (*Terminalia arjuna*) bark in Central India, and the endangered Sarpagandha (*Rauwolfia serpentine*). These high-demand medicinal plants were revived from their threatened status through regulation of collection, improvement in collection practices, and large-scale planting and cultivation. This was followed up with setting protocols of harvest and grading product quality. A discerning and regulated market of these products also played a role, providing good rates to the producers and setting standards of quality gradation.

Many actors came together for this intervention, with government research institutions, state Forest Department, NGOs, and local farmers engaged in the revival and subsequent conservation of these medicinal plants. A practice currently used to ensure sustainability and meet market demand is to augment production of NTFP by cultivation (Figure 5.6, Figure 5.7). This method uses simple market-based mechanisms of demand and supply. When the supply of a natural resource from the forest is exhausted, the resource is taken up for cultivation. One such effort involving poor tribal communities has enabled both revival of the NTFP and generates significant additional income for cultivator families.

Nannari (*Decalepis hamiltonii*), which is called 'Maredu Gaddalu' in Telugu, is a high income NTFP for Chenchu and Yanadi communities in Rayalaseema Region of Andhra Pradesh. Its root is used in herbal medicines and pickles and to make a locally popular health drink. Girijan Cooperative Corporation has been procuring this root for the past 20 years and has introduced a new value-added product, 'Nannari Sharbat' which has proved extremely popular with consumers. The popularity of Decalepis in the national market has recently made its price soar. There is a severe pressure on this NTFP, and apprehension that this plant may become endangered due to over-exploitation.

In this context, between 2014 and 2015, Kovel Foundation, in consultation with the Forest Department, tried to cultivate Nannari by purchasing

*Figure 5.6* Medicinal plants raised in a nursery

*Photo: Kovel Foundation*

*Figure 5.7* Medicinal plant nursery

*Photo: Kovel Foundation*

*Figure 5.8* Farmers with root of *Decalepis hamiltonii*

*Photo: Kovel Foundation*

6,000 saplings and giving them to six farmers in Kurnool District (Figure 5.8). Arogyaiah, who belongs to the Chenchu community and who duly followed the prescribed 'Package of Practices', was able to produce 300 kg of dry Nannari root. Selling at Rs. 200 per kg, it gave him a total earning of Rs 60,000. This was a big success among the community; especially since such an income was generated from as little as 0.25 acres of land.

Arogyaiah's inspiration led to other farmers taking up Nannari cultivation, and Kovel has been involved in organizing a series of meetings to highlight and popularize Nannari cultivation in the region. There are other farmers who have earned between Rs 1 lakh and 1.6 lakhs from just 0.15 to 0.20 acres of land. In the process of making this a community-centric activity, Kovel has motivated and technically equipped farmers to raise nurseries themselves. They have successfully completed plantation of around 1 lakh seedlings raised as an inter-crop, border crop, and pure crop following natural farming methods. By conservative estimates, these seedlings should be able to generate around Rs. 2 crores as income in a period of two years.

This intervention has great potential, which can change local livelihoods of the Chenchu community, and help to protect and conserve the plants in the forest from overharvesting. Keeping the high incidence of landless families in the region in view, Kovel has designed an innovative model namely 'Chadaram Model' for raising 100 Nannari plants next to leafy vegetables.

The design involves a raised soil bed (2 feet), with an area of 12 × 20 feet, fenced in with stone plates (1 meter), to facilitate effective management of cultivation and efficient harvesting of the roots. This model seems to have the potential to generate an income of about Rs. 40,000 in 24 months. A total of 20 landless Chenchu farmers have taken this up. Kovel Foundation, in collaboration with Quality Counsel of India (QCI) & National Medicinal Plants Board (NMPB) in New Delhi, has developed a strategy for listing Good Agricultural Practices (GAP) for Nannari cultivation and is also engaging with certification to ensure premium prices for farmers.

### *Western Ghats, India – interventions by Keystone Foundation for sustainable NTFP management*

Keystone Foundation has made a concerted effort, over 27 years, with the harvesting of *Apis dorsata* wild honey (Figure 5.9) in the Nilgiri Biosphere Reserve that lies at the trijunction of Kerala, Karnataka, and Tamil Nadu. The reserve has at least seven indigenous communities who practise honey collection seasonally. Traditional practices with complex cultural dimensions are a part of honey hunting. The knowledge of forest vines, flowering, bee behaviour, and comb building seasons is intrinsically understood by these communities. Earlier used for barter, harvested honey now fetches a good market price.

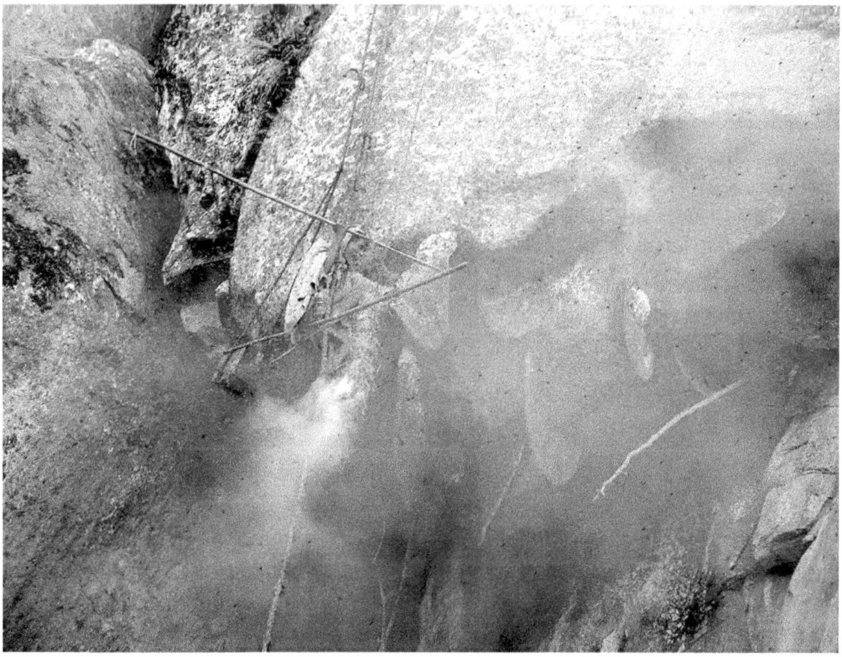

*Figure 5.9* Honey collection from combs of *Apis dorsata* on Nilgiri cliffs

A field survey done in 1994 by the founders of the organization while backpacking across the state of Tamil Nadu, highlighted the fact that the quality of honey was poor due to unhygienic practices followed at harvest and post-harvest stages. The economic returns to the community were also low due to an unorganized market. Since then, the organization has attempted to address this issue with a multi-dimensional approach. The focus of the work being done over the years is to ensure ecologically sound practices to address issues of sustainable livelihoods of forest-based communities, and biodiversity conservation of the mountainous region, drawing on the strength of the local communities' experiences with wild honey collection.

Local communities are involved in every facet of this work, which hinges on their participation and consequent empowerment. Ecological activities and field research are carried out by community members trained as 'barefoot ecologists' who also do the monitoring work. Facilities for laboratory testing were set up in the local area, and field-testing kits are now used to explain the results to the people. Similarly, benefits to the community were assured by fixing quality parameters and higher returns. It took a few years to wean people away from the clutches of the traders as they began to see the benefits of people-based institutions. The setting up of Thumbitakadu and Thoduve as cooperatives and the Farmers Producer Organisation – Aadhimalai gave the power of marketing and profit sharing to the community. Similarly, local value addition by indigenous women has resulted in empowering communities with both knowledge and natural resource rights.

With honey hunting, sustainability of harvest and protection of habitat is ensured by the members of the indigenous communities, who also have the option to retain their traditional lifestyles, should they choose to. This example has empowered communities in NTFP management, emphasizing the importance for them to understand the links between the economic and the ecological aspects of the trade. The model is driven by the three pillars of sustainability (Figure 5.10) and is made more effective by the use of research and appropriate technology, as well as monitoring and adaptation to changing times.

## Lessons learned and challenges faced for sustainable NTFP management

Key lessons that have been the focus of this chapter are listed as follows:

- Commercially harvested NTFPs need to be monitored regularly and data analysed to see what drives unsustainable practices. In the case of Tendu Leaf, insufficient data and context have made it difficult to ascertain the explanation for a declining trend of collection.
- From the case of Gum Karaya, the collaboration of institutions emerges as a key element. The example is important to understand how researchers and scientists can work with NGOs and marketing institutions to

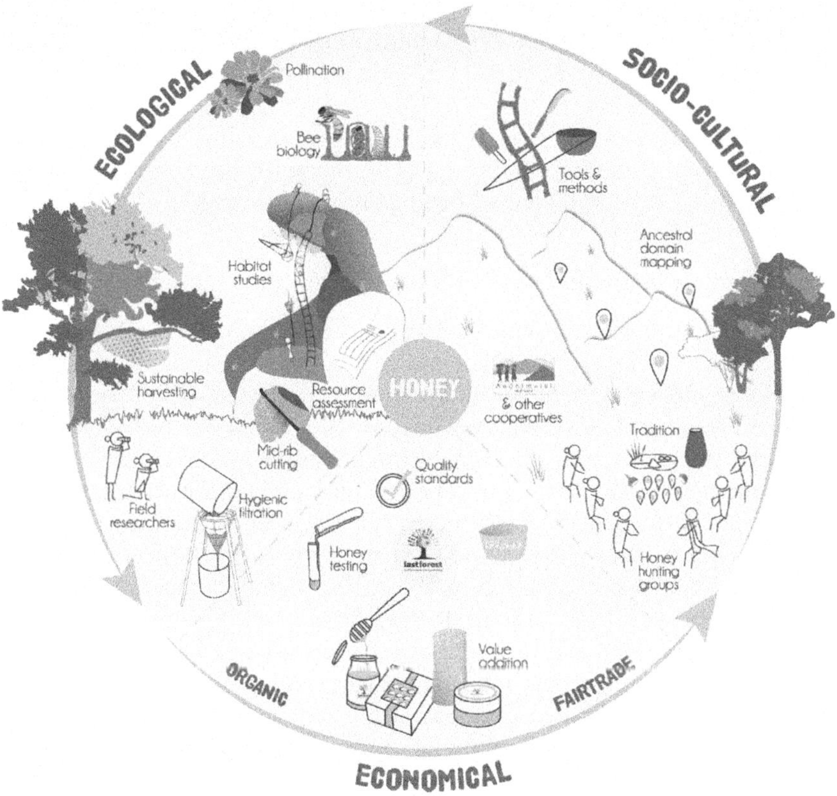


*Figure 5.10* Ecological, socio-cultural, and economic approaches used in the sustainable management of wild honey

*Illustration: Udaya Vauhini*

come up with sustainable solutions. However, this collaboration may be difficult to replicate in every circumstance.

- From the case of honey harvesting, we can see concerted long-term work on several dimensions of the NTFP using a community-based approach. Here too, the role of the NGO, which draws on external knowledge and indigenous knowledge, is important. Keystone Foundation has come up with hybrid solutions for sustainable harvesting and livelihoods, stemming from its partnership with communities. This relationship-building coupled with technical input may be difficult to replicate across the range of commercially harvested NTFPs. Nevertheless, there is scope for experimentation along these lines.
- From the case of cultivating Nannari in small patches of land, it appears that it is important to select species of high value so that returns are

maximized. As a two-year crop, it is possible that the care and maintenance of Nannari takes more effort, which is difficult to account for. On the other hand, root stock is often taken from the forest, which can further endanger the species. When cultivation reaches a stage of providing enough for both seed and sale, the process will be more sustainable.

Overall, it is evident that several factors impact the sustainability of commercial NTFPs. Constant monitoring and community engagement is critical to recognizing and addressing all the dimensions involved.

## Strengthening sustainable use practices in commercial NTFPs

The most important factor which can strengthen sustainable use is the implementation of policies and laws that ensure tenure to tribal communities. The opposite is unfolding today with development projects and schemes reducing forest cover and destroying the basis of NTFPs and related livelihoods. The poor implementation of FRA has deprived people from owning and managing their forests, including NTFPs. Policy and regulatory frameworks will have to look at NTFPs in a nuanced way, keeping ecological, social, and economic aspects in its purview. There is no specific policy on NTFPs in India as this issue is overseen by several Acts and administrative orders under guidelines such as the Joint Forest Management Guidelines, 1990.

These development and transactional aspects need to be seen in light of the resource base. Regeneration, plantations, and conservation of threatened species need to be increased. Forestry and plantation projects (e.g. The Compensatory Afforestation Fund Management and Planning Authority) do not take NTFPs into consideration when selecting species and prefer timber monocultures. An analysis related to availability, ownership, benefit sharing, monopoly rules, tax, and transit rules surrounding NTFP collection and trade is much needed.

## Conclusion

This chapter outlined examples from the field that reveal that long-term efforts are necessary to foster sustainable use of NTFP, given the complex and dynamic nature of factors that determine such use. Schemes that benefit people through product quality standards, market information, value addition, and employment will bring sustainable practices to the forefront and lead to community-based monitoring that will contribute to biodiversity conservation in the long run.

Certification of NTFPs and building sustainable markets is important for strengthening sustainable use among commercially harvested NTFPs. Most certification methods developed for NTFPs follow expensive third-party standards and are unaffordable for many. However, certification processes follow important ecological, social, cultural, and economic parameters ensuring sustainability and fair market dealings. Relatively inexpensive, participatory, and community-based certification has yet to be developed for NTFPs, though models of the same are prevalent in the case of agricultural products.

Importantly, there is a paucity of research on sustainable harvesting of NTFPs. This type of research needs to be encouraged through institutions across the country. Lack of collaborative research and participatory approaches have created silos of knowledge. This has also meant that research and knowledge are not driving policies. Development initiatives also bring forth short-term solutions only focused on income, often degrading the very resource on which it depends. Interventions, such as those showcased in this chapter, highlight that strategies that incorporate people-based solutions and enable community management of NTFP will have long-term application for poverty alleviation and conservation. Important change can be brought by increasing collaboration between researchers and practitioners, and people and policymakers to engender shifts in the development model that can work towards empowered societies.

## References

Bhattacharya, P., & Hayat, S. F. (2009). Sustainable NTFP management for livelihood and income generation of tribal communities: A case from Madhya Pradesh, India. In R. Uma Shaanker, A. J. Hiremath, G. C. Joseph, & N. D. Rai (Eds.), *Non timber forest products: Conservation management and policy in the tropics* (pp. 21–34). Bangalore: ATREE & University of Agriculture Science.

Harilal, M. S. (2009). Commercialising traditional medicine: Ayurvedic manufacturing in Kerala. *Economic & Political Weekly*, *44*(16), 44–51.

Krishnamurthy, V., Mandle, L., Ticktin, T., Ganesan, R., Saneesh, C. S., & Varghese, A. (2013). Conservation status and effects of harvest on an endemic multipurpose cycad, Cycas circinalis L., Western Ghats, India. *Tropical Ecology*, *54*(3), 309–320.

Lele, S., Pattanaik, M., & Rai, N. D. (2010). NTFPs in India: Rhetoric and reality. In S. A. Laird, R. J. McLain, & R. P. Wynberg (Eds.), *Wild product governance: Finding policies that work for non-timber forest products* (pp. 85–113). London: Earthscan. https://doi.org/10.4324/9781849775199

Malhotra, K. C., & Bhattacharya, P. (2010). *Forest and livelihood: Status paper (Forest ecosystem)*. Hyderabad, India: Research Unit for Livelihoods and Natural Resources, Centre for Economic and Social Studies.

Mansourian, S., Vira, B., & Wildburger, C. (Eds.). (2015). *Forests and food: Addressing hunger and nutrition across sustainable landscapes* (pp. 288). Open Book Publishers. DOI: 10.11647/OBP.0085

Pain, A. (2009). What is driving change in the Nilgiri Biosphere Reserve and what effects might such change have on the role of NTFP in the livelihoods of indigenous people? In R. Dutt, J. Seeley, & P. Roy (Eds.), *Proceedings of the biodiversity and livelihoods conference*. Kotagiri, India: Keystone Foundation.

Poffenberger, M. (1994). The resurgence of community forest management in Eastern India. In D. Western, R. M. Wright, & S. C. Strum (Eds.), *Natural connections:*

*Perspectives in community-based conservation* (pp. 53–79). Covelo, CA: Island Press.

Setty, R. S., Bawa, K., Ticktin, T. and Gowda, C. M. (2008). Evaluation of a participatory resource monitoring system for nontimber forest products: The case of amla (Phyllanthus spp.) fruit harvest by Soligas in South India. *Ecology and Society*, *13*(2), 19.

Shackleton, C., Shackleton, S., & Shanley, P. (2011). Building a holistic picture: An integrative analysis of current and future prospects for non-timber forest products in a changing world. In S. Shackleton, C. Shackleton, & P. Shanley (Eds.), *Non-timber forest products in the global context*, *Tropical forestry book series volume 7* (pp. 255–280). Berlin, Heidelberg: Springer. https://doi.org/10.1007/978-3-642-17983-9_12

Shackleton, C. M., Pandey, A. K., & Ticktin, T. (2015). Ecologically sustainable harvesting of non-timber forest products: Disarming the narrative and the complexity. In C. M. Shackleton, A. K. Pandey, & T. Ticktin (Eds.), *Ecological sustainability for non-timber forest products* (pp. 274–292). New York: Routledge.

Shanker, K., Hiremath, A., & Bawa, K. (2005). Linking biodiversity conservation and livelihoods in India. *PLoS Biology*, *3*(11), e394.

Ticktin, T. (2004). The ecological implications of harvesting non-timber forest products. *Journal of Applied Ecology*, *41*(1), 11–21. https://doi.org/10.1111/j.1365-2664.2004.00859.x

Ticktin, T., & Shackleton, C. (2011). Harvesting non-timber forest products sustainably: Opportunities and challenges. In S. Shackleton, C. Shackleton, & P. Shanley (Eds.), *Non-timber forest products in the global context*, *Tropical forestry book series volume 7* (pp. 149–169). Berlin, Heidelberg: Springer. DOI:10.1007/978-3-642-17983-9_7

# 6 Sustainable use of wild medicinal plant resources

## Developing field methods for sustainable collection and direct market linkages

*Jagannatha Rao R. and Deepa G.B.*

### Introduction

Sustainable use is an effective tool to protect and conserve biological resources. The basic idea behind sustainable harvesting is that a biological resource should be harvested within the limits of its capacity for self-renewal. Furthermore, the manner of its harvest should not degrade the environment in any way (Hamilton 2005). Sustainable harvesting is the use of plant resources in such a way that the plants are able to continue to supply indefinitely (Wong et al. 2001). This harvesting method places an emphasis on maintenance of species populations in the wild, irrespective of high demand.

In the absence of any guidelines and protocols, collectors might simply go and gather all available plant material or the quantity demanded by the trader. Nevertheless, in some regions, collectors, especially tribes and forest dwellers, still follow collection methods passed on to them as a tradition or way of life that also happen to be sustainable. Such stewardship is threatened when the market does not provide an incentive for sustainable collection. For this reason, wild collection guidelines and protocols need to be developed and promoted.

There have been initiatives in the past at the international level to develop sustainable wild collection standards and guidelines. These attempts were primarily aimed to ensure sustainable wild collection of raw drugs in compliance with the guidelines that once certified, would be preferred by industries. Despite pilot efforts around the world and subsequent ratification of standards, these international guidelines for implementing sustainable wild collection, being too broad, have since become irrelevant among stakeholders.

### Developing wild collection guidelines

Any attempts to develop wild collection guidelines and protocols should consider three aspects for successful implementation by collectors – an integration of traditional collection practices, scientific species-specific recommendations on wild collection, and flexibility to make amendments based

DOI: 10.4324/9781003343493-8

on lessons learned. Traditional collection methods, despite being practiced at small scales for years, are time-tested for their sustainability factor. In addition to giving due recognition to age-old wisdom on collection practices, the merging of old and modern knowledge systems make these guidelines widely acceptable among stakeholders, especially local communities of collectors.

Trans Disciplinary University (TDU) promoted by the Foundation for Revitalization of Local Health Traditions (FRLHT), collaborated with international, national, and regional agencies to develop principles and practices of sustainable harvesting of wild species. Through these efforts, TDU has developed species-specific sustainable wild collection protocols for 48 species at diverse geographical locations in India. The lessons learned during the field implementation of sustainable harvesting methods were consolidated and a five-phase implementation practice evolved from these lessons. This resulted in the production of the Practical Guidelines for Field Implementation of Sustainable Wild Collection of Medicinal Plants in India Guidelines.

The Guidelines contain general and species-specific protocols. General guidelines developed for harvesting of plant parts provide for sustainable management of resources in the unit area, which contributes towards regeneration of plant species and ensures supply of genuine and quality plant material by adopting community-based post-harvesting operations. Species-specific guidelines provide information on harvesting protocols, phenological patterns, regeneration ability, pollinators and dispersers, and the likely impacts of harvest, all of which contribute to ecological and economic sustainability.

Field implementation of guidelines was undertaken in collaboration with state Forest Departments and community-based organizations, which contributed to revisions of the Guidelines on account of harvest impacts over time, demand fluctuations, and the practical difficulties of implementation. The Guidelines cover the following aspects:

- Sustainable collection: Protocols for resource inventory, wild collection, and value addition have been developed for species that are heavily traded or of conservation concern. Species-specific resource inventory and assessment were developed with vegetation mapping to understand the stock of prioritized medicinal plant species in the harvesting areas. Additionally, a quantity assessment tool was formulated, to be deployed prior to collection, in order to measure harvestable plant materials available for each harvesting season from specific sites.
- Good collection practices: Collection of plant materials is permitted only after conducting the quantity assessment. Species-specific protocols on good collection practices (GCP) have been developed to maintain the

resources in the habitat long term. Questions covered in the GCP protocols are as follows:

- what to collect?
- when to collect (season and period)?
- what stage of plant material to collect?
- how to collect?
- how much to collect?
- how to transfer?

- Value addition practices: After plant materials are collected from the wild, they are brought to a procurement centre and registered there. Simple value addition practices are employed to deliver relevant plant products for industry use. Guidelines for value addition of wild-collected plant materials provide detailed information on maintenance of collection registers, semi-processing methods such as drying and powdering, product quality checking for adulterants, and the storage, packing, labelling, and transporting of plant materials.

Institutional mechanisms have been developed for planning, collection, post-harvesting, value addition, and marketing of sustainably collected material. Each primary collector is identified and given an identity card, which streamlines participation and ensures measurable economic and social benefits.

## Principles of sustainable harvesting

TDU-FRLHT's experience in working with local communities and other stakeholders has culminated in the development of guiding principles for sustainable collection of medicinal plant species that relies on an ecosystem approach and suggests landscape-level management to promote species-specific conservation and sustainable use.

### I: Resource mapping, estimation, and value analyses

This principle enables local communities to understand the availability of resources, potential quantity that can be harvested, their conditions in the wild and impacts of destructive harvesting. Furthermore, it allows participating stakeholders to understand the value of products harvested in economic terms and the conservation of resources.

### II: Participatory approach

The participatory approach integrates people of different socio-economic status to establish need-based and objective-oriented local institutions, which can specify roles and responsibilities, demarcate their dependency on the resource, and assess their contribution towards the conservation and sustainable use of wild resources.

### III: Assimilation and application of traditional harvesting knowledge

Documentation of indigenous knowledge related to phenology, distribution, climate and productivity, animal interaction, regeneration, propagation, cultural, and spiritual relevance helps with building local species profiles, their medicinal values and traditional uses, harvesting patterns and processing. These have to be combined with academic knowledge to co-produce sustainable harvesting protocols. This principle prescribes dissemination and capacity building programmes for the different stakeholder groups involved in the process. They are designed to develop and enhance the adaption of sustainable harvesting methodology.

### IV: Sustainable harvesting protocols, tools, and field implementation

This principle works to create species-specific and location-specific sustainable harvesting best practices by combining traditional and academic knowledge. These will be implemented in the field by establishing institutional mechanisms that ensure these are followed.

### V: Training, capacity building and dissemination

This principle prescribes dissemination and capacity building programmes for the different stakeholder groups involved in the process. They are designed to develop and enhance the adaption of sustainable harvesting methodology.

### VI: Post-harvesting techniques and market linkages

Post-harvesting techniques such as drying, storage, value addition, labelling, and branding fetch higher prices in the market and also generate additional income and employment for locals. The resulting value-added products are sent directly to the industries to ensure supply of quality raw material. Marketing is an important component of sustainably harvested products, as this is the incentive that compensates sustainable collection practices.

**VII: Linking trade and conservation**

This principle integrates stakeholders that have different roles in the value chain of resource collection, value addition, and marketing. The idea is to directly link the primary stakeholders from the resource base (collectors and village level institutions) with the marketing and trade of sustainably collected resources by assimilating possible value addition technologies at the local level. This will generate additional income and employment, thereby encouraging and enhancing the participation of stakeholders in the conservation of resources.

## Case study: linking raw drug plant material with herbal industries in Kerala, India

The National Medicinal Plant Board, Ministry of AYUSH, Government of India awarded a project to the Silent Valley Forest Development Agency and the Peechi Forest Development Agency of the Kerala Forest & Wildlife Department to support nine eco-development committees (EDCs). These committees would promote the sustainable harvesting of wild medicinal plants by establishing infrastructure facilities for storage, grading, drying, post-harvesting, and value addition, as well as marketing value-added products. The goal of the project was to develop sustainable institutions and build the capacity of EDCs to handle wild medicinal plant resources, in order to generate additional employment and income for marginal groups.

TDU-FRLHT designed the project and provided technical support to the implementing agency, that is the Kerala Forests & Wildlife Department. The project aimed to deploy the Guidelines in a field setting, through regular training and capacity building of local collectors. Additionally, the collectors, especially women and youths, were supported to perform value additions locally, and provided support with market linkage to trade their collections for a good price.

### *Area profile*

The selected EDCs in Silent Valley National Park and Peechi are located in the buffer zones of protected areas that are rich in natural resources, especially medicinal plants. Tribal communities in these areas, who are members of the EDCs, completely depend on these resources for their health and livelihood security. Medicinal plants contribute substantial income to tribal communities in the selected EDCs. They are also important in terms of their medicinal uses, both locally and at the national level, where they are highly traded commodities. The selected EDCs are located in colonies where only schedule tribe communities live, and there are no other livelihood modes available other than some farming, which is undertaken at subsistence scale.

### *Existing supply value chain*

The Silent Valley National Park, Mukkali and the Peechi Wildlife Office, Peechi have a medicinal plant collection centre and a 'Vanasree' eco-shop. However, there are currently no infrastructure facilities for manufacturing products or for marketing and trading these plants available to the EDCs. Plant species are locally collected and carried to the nearby towns. Middlemen – local traders and agents, purchase raw material from the collectors at low prices. There are a few traditional healers who collect plant materials from forests through hired labour. However, no information is available on their consumption. Medicinal plants are also used by village elders.

### *Baseline survey and selection of species*

Baseline data on medicinal plants were collected from each EDC. It included the number of active collectors, medicinal plants and parts collected, quantity collected, price offered, and place of marketing. With this information, 14 medical plant species and honey were prioritized for developing sustainable harvesting protocols. These included species under the IUCN red list categories – vulnerable (4), near threatened (2), and endangered (1).

### *Setting up an institutional framework*

Attempts were made to strengthen EDCs through creating awareness about medicinal plant resources in the wild. A task team was formed for each EDC, to conduct participatory exercises to collect information on species and collection practices. Stakeholders included the Forest Department, traditional healers, women's groups, plant collectors, traders, and local government members. This team of 10 to 15 members functions as a subsidiary body under the EDC, which is an existing institution established by the Forest Department. Medicinal plant collectors in each EDC were identified and their personal details collected, along with photographs. A total of 473 collectors were thereby issued identity cards, with unique codes, with the intention of monitoring the movement of plant materials from collection sites and the quantity collected from the wild.

### *Infrastructure and capacity building*

Infrastructural facilities including storage areas, drying sheds, processing equipment, and collection tools were planned for each EDC. A detailed assessment of possibilities of value addition options for prioritized species was prepared. To initiate processing activities, the list of tools and equipment required for the value addition process, details of the design and cost of tools, and list of vendors was prepared. The local management task team was trained on GCP and value addition, through interactive meetings, field training programmes, and exposure visits. Training and communication materials such as posters, presentations, manuals, and pamphlets were prepared to support the task team. Trained team members in turn are expected to train and equip other plant collectors on GCP during wild collection.

### *Market linkages*

Market linkages between plant collectors and pharmaceutical companies were initiated through organizing regular meetings between industries and collectors who are part of EDCs. Prior to this, a list of potential Ayurveda firms that would be interested in procuring sustainably harvested and value-added products of the prioritized species was prepared. Buyers of raw drugs were visited and oriented on the project's interventions. They were informed that the meetings would be avenues to discuss the modalities of direct procurement of raw material from tribal collectors. Arrangements are underway for MoUs between pharmaceutical companies and EDCs, in order for the former to participate in the auction of wild-collected plant materials. Figure 6.1 and Figure 6.2 show market linkages before and after the project.

A revolving fund has been arranged to offer cash to collectors when they bring plant materials to the procurement centre. Each task team has a dedicated sub-team, comprised of women and youths, who are responsible for managing the funds and maintaining records.

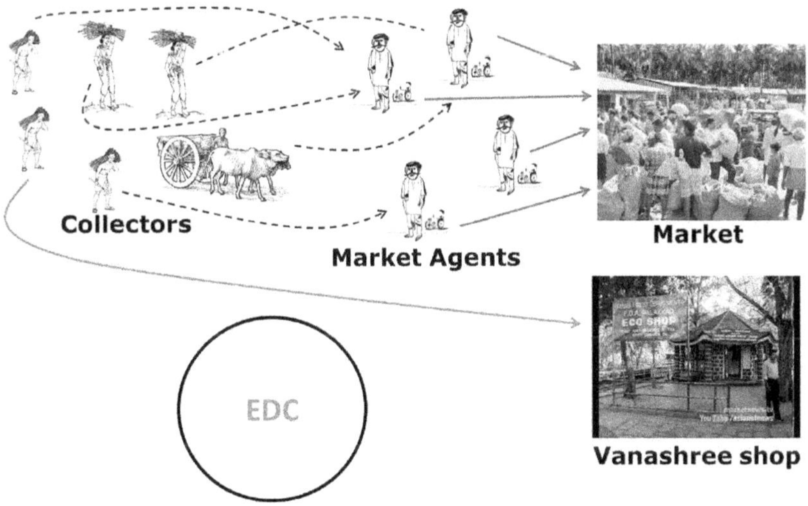


*Figure 6.1* Wild plant market channels before intervention

Source: *Rao and Selwyn (2021) Forest-plus 2.0 Report*

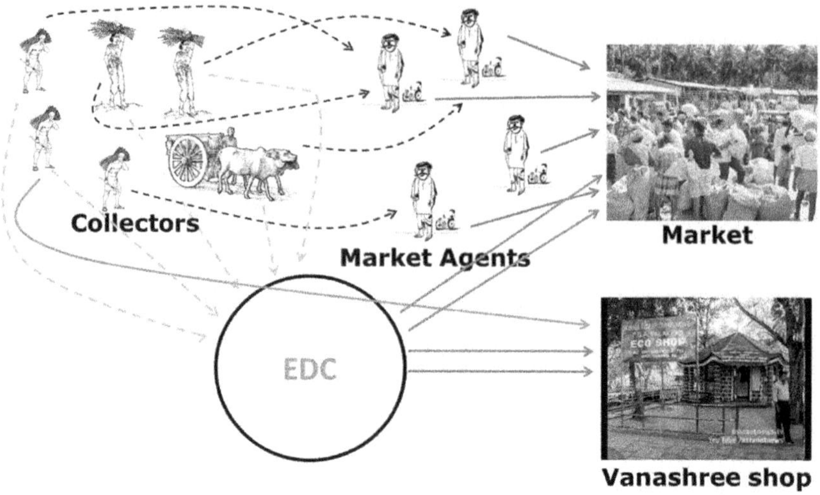


*Figure 6.2* Wild plant market channels after intervention

Source: *Rao and Selwyn (2021) Forest-plus 2.0 Report*

## Lessons learned and challenges faced

Plant collectors in EDCs are now aware about the medicinal use of their collection and their significance in herbal industries and the manufacturing of medicines. Despite the successes of the project, the following emerge as challenges in fostering participatory sustainable harvesting protocols and their practices:

- Large numbers of species are ecologically threatened and traded in high volumes with differing diversity, ecology, habitats, and pressures. This poses serious problems for maintaining biological diversity in the face of wild collection.
- The need to develop GCP is specific to species and geographical location. Interventions therefore cannot be duplicated, but need to be updated to suit the needs of each site.
- There are unpredictable and often unrecognized environmental factors that influence annual yield of wild populations.
- Unclear land and resource tenure, and uncertain administrative authority operating in varied locations.
- The sheer vastness and variety of products, uses, and markets.
- Long, complex supply chain between source and market in most areas.
- Lack of proper post-harvesting techniques for agriculture produce.

- Absence of customer awareness on the quality and harvest of the wild plants contained in products they consume.

Additionally, small-scale interventions such as described earlier occur sporadically. There continues to be limited recognition of the economic, social, and ecological value of wild resources, and a widespread uncertainty about who is responsible for ensuring that wild resources are used sustainably.

## Conclusion

Unscientific collection from the wild has led to the threat of extinction and severe genetic loss in the wild. There is an urgent need to create a strategy for appropriate management of medicinal plants growing under widely varying habitat conditions in the country, in a way that meets local needs and conserves biodiversity. Complex systems of sustainable harvesting, processing, and trade have been hard to understand without standardized methods and models. This has been a barrier to introducing transparency in marketing and quality control. As this chapter set out, sustainable harvest methodologies hinge on local knowledge, secure tenure rights, and autonomy of local institutions. Institutional capacity, equity and resource access, market policy, and trade linkages need to be addressed to develop an adaptive operational management strategy.

Support is needed from policymakers to incorporate the principles and practices of sustainable use of biodiversity components under the provisions of the Biological Diversity Act. Policy interventions are needed to incentivize such sustainable use, accompanied by mechanisms for monitoring, auditing, and certification. EDCs and Biodiversity Management Committees are ideally placed to undertake sustainable resource management, while gaining from benefit-sharing mechanisms. Creating local autonomy at this scale can lead to effective decision making.

## References

Hamilton, A. (2005, July). Resource assessment for sustainable harvesting of medicinal plants. In a side-event at the International Botanical Congress on Source to Shelf: Sustainable Supply Chain Management of Medicinal and Aromatic Plants, Vienna (pp. 21–22).

Wong, J. L. G., Thornber, K., & Baker, N. (2001). *Resource assessment of non-wood forest products: Experience and biometric principles, Non-wood forest products series 13*. Rome: FAO.

# Part III

# Community knowledge

# 7 The pig and the turtle

## An ecological reading of ritual and taboo in ethnographic accounts on Andamanese hunter-gatherers

*Meera Anna Oommen*

### Introduction

Human relationships with the natural environment are often guided by informal institutions consisting of rules, norms, and prohibitions that are decided upon autonomously by traditional societies. Termed 'resource and habitat taboos' (RHT), these are the result of long-term engagements of a society in a landscape (Colding and Folke 2001) and are meant to be adaptive, that is 'anything that increases the probability of survival of the individual or the group' (Reichel-Dolmatoff 1976, p. 308).

On account of their development across several interrelated axes, it is difficult to separate their origins, and their functions often serve overlapping social, ecological, and psychological ends (Gadgil and Guha 1993, Colding and Folke 2001). In recent years, their significance has been explored for a range of human and environmental well-being outcomes – these include locally derived protection measures for human health and nutrition as well as that for the environment (e.g. tackling climate change, ensuring the long-term sustainability of species and habitats).

Early studies that explored informal institutions relating to ecological norms, carrying capacities, and sustainability include Rappaport (1968), Harris (1971), Reichel-Dolmatoff (1976), and Johannes (1978, 1981). These pointed to the effectiveness of adaptive responses embedded in local rules and cosmologies relating not only to ensuring sustainability of species and natural resources but also related to interlinked regulation of human populations and behaviours via ritual and taboo (Reichel-Dolmatoff 1976). More recent work has investigated the potential of such informal systems to relate to modern conservation and sustainable use and synthesized information related to different types of RHTs.

### Resource and habitat taboos

Colding and Folke (2001) identify six broad categories of RHTs which are of significance for conservation and are applicable across traditional systems in different contexts. These include segment taboos (which regulate

DOI: 10.4324/9781003343493-10

the removal of a species or resource for a certain purpose, e.g. species-specific food taboos), temporal taboos (resources are regulated for a period of time, seasons, etc.), life-history taboos (use of a species during a certain life stage, especially vulnerable states are avoided), species-specific taboos (the removal of certain species are prohibited), habitat taboos (certain habitats are off-limits), and method taboos (some types of hunting and harvesting gear and types of capture are prohibited). Often, RHTs are of an overlapping nature and act in conjunction with each other.

With its wide array of biogeographic units and diverse local communities, the Indian subcontinent has had a long history of use of natural resources, as well as traditional systems of management. These include informal institutions that rely on social taboos and norms. However, in the contemporary Indian environmental protection scenario, the compatibility of traditional use with modern conservation strategies has been poorly adjusted for several reasons. Exclusionary strategies of colonial and princely governments followed by independent India's fortress conservation strategies, starting with the *Wildlife (Protection) Act, 1972*, enclosed wild spaces and excluded local communities from protected areas. This highly preservationist approach has been the cornerstone of environmental management spearheaded by the state.

This was also supported by the non-governmental conservationist lobby (largely dominated by the urban elite), with little appreciation for the use of species as a conservation strategy, as opposed to their complete protection. As a consequence, there has been limited recognition of the potential for strategies that involve coexistence and use. Furthermore, there has been an attempt at 'reform' that involves the active removal of forest-dwelling and ecosystem-dependent communities by way of relocation, sedentarization, and gentrification that have in turn resulted in the loss of ties to the land, culture, and even identity – a favourite being turning erstwhile hunters into protectors of species that were once utilized. In this regard, there are parallels with the global conservation discourse. These include recent calls that caution conservationists against 'moral relativism' and 'misguided respect for cultural backgrounds' and exhortations to take the side of conservation in 'culture–conservation conflicts' in less developed countries (Dickman et al. 2015).

An exception to the prevalent model of protectionist conservation in India can be found in some parts of the mainland (e.g. the north-east region of India), as well as the Andaman and Nicobar Islands, where traditional systems of livelihoods that involve practices such as hunting of wildlife and harvesting of wild products are sanctioned in the case of indigenous communities. This chapter aims to understand social practices within such systems to identify the potential contribution of informal rules and taboos to sustainable use and conservation in one such geography, the Andaman Islands. This is explored through a review of literature that includes ethnographic work in colonial accounts and more recent investigations by anthropologists.

The primary anthropological sources of the colonial era include early accounts by Mouat (1863), Man (1883), and Portman (1899) and followed by a more detailed ethnographic attempt by Radcliffe-Brown (1922). Anthropological sources post-Indian independence include work by Cipriani (1966) and Cooper (1990), and a comprehensive body of anthropological scholarship on the Ongee in Little Andaman by Pandya (1993, 2009). Even though the question of ecological sustainability was not a primary focus of these analyses, a close reading of this literature reveals a range of insights about Andamanese communities from the perspective of local management and sustainable use of resources in island systems with finite resources and constraints on space.

Furthermore, it needs to be stressed that our own analysis of ritual and taboo is different from the islanders' own sensibilities. Our compartmentalization of these instruments as serving primarily ecological and other functional ends is unlikely to be viewed in the same way by the islanders for whom these form constituents of a holistic cosmology. As ecological sustainability is only one axis of everyday engagement between indigenous communities and their environment, it is imperative that this distilling of RHTs be read along with work such as by Pandya (1993, 2009) for a comprehensive understanding of these interrelated components of local cosmologies.

## Andaman and Nicobar Islands

Situated along the north-western extremity of Island South East Asia (ISEA), the Andaman Islands form a part of the Indo-Burma global biodiversity hotspot. They host a diverse, yet distinct set of biogeographical features that set them apart from the Indian mainland. While most of the hunter-gatherer communities such as the Great Andamanese groups were decimated by colonial contact or have been assimilated into mainstream culture through various 'civilizing' missions, groups such as the Jarawa on South and Middle Andaman continue to lead somewhat secluded, natural resource-oriented lifestyles aided by the prevalent policy of limited contact. The Ongee of Little Andaman are primarily segregated within reservations. The islanders of North Sentinel remain the only indigenous group that resides without formal contact with the outside world. Their seclusion is currently facilitated by the Indian government's strict no-contact policy.

The Andaman Islands are also home to settler communities from different regions on the Indian mainland that vastly outnumber the indigenous groups on most islands. They include local-born descendants of erstwhile convict settlers from mainland India, settler groups brought into the islands from various parts of the Indian subcontinent for a variety of reasons including as labourers, refugees, and people displaced due to political unrest (e.g. Karen, Moplah, Ranchi, Bhatu, Bengali, Telugu, and Tamil). Altogether, these groups constitute a unique, socially complex assemblage of communities. Due to their unique biogeographic position in the Indian Ocean, and their histories of occupation, these islands are centres of biocultural diversity, making them significant sites for conservation not only for terrestrial and marine species and ecosystems but also that of cultural complexes related to the use of species, local traditions, and indigenous cosmologies.

## The pig and the turtle

Despite their problematic racial stereotyping of indigenous communities, accounts from the colonial period remain among the few sources of historical information on Andamanese communities. As is typical of colonial misconceptions about the tropics, the lack of agriculture and the translocationary and hunting ways of life of the islands' forager groups were viewed contemptuously by early colonial officers. Lt. Colebrooke (Colebrooke 1799, p. 407), the author of one of the first accounts, calls their mode of life degrading to human nature. Later, observers such as Mouat (1863), Man (1883), Portman (1899), and Radcliffe-Brown (1922) were equally committed to the notion that the savage islanders were indeed in need of civilization and Christianity. In spite of this general disdain, the hostility of the islanders to outsiders (their most marked particularity) and their 'brutish' lifestyles aroused significant ethnographic interest.

Colonial ethnographic accounts grouped Andaman islanders within two main divisions: the Great Andaman Group and Little Andaman Group respectively. The Great Andaman Group included the inhabitants of the Great Andaman – North, Middle, and South Andaman and surrounding islands – with the exception of the Jarawa on South Andaman. They were further divided into 10 ethnolinguistic subgroups (*Aka Cari, Aka Kora, Aka Bo, Aka Jeru, Aka Kede, Aka Kol, Oko Juwoi, A Pucikwar, Akar Bele, and Aka Bea*). The few remaining individuals of these groups are now restricted to Strait Island. The Little Andaman Group – once all referred to as Jarawa – included the inhabitants of the Little Andaman (Ongee), North Sentinel and the Jarawa of the South Andaman. Currently, four major groups remain, collectively referred to as the Andamanese: the Ongee, the Sentinelese, the Jarawa, and the remaining individuals of Great Andamanese (restricted to Strait Island).

### *Great Andaman*

Based on their hunting lifestyles and geographies, the Andamanese themselves were known to group themselves into the forest dwellers and the coast dwellers. According to Man (1883), in the *Aka Bea* language, the *ē.remtâ.ga* (the forest dwellers living inland) were known to address an *àryô.to* (coast-dweller) as *ēr-châ.tâknga* (one who loses his way) or *ēr-lō.inga-ba* (one who cannot find his way in the jungle), or *un-päg-ī.kng-ba* (one who cannot follow tracks [of pigs]).

The *àryô.to* in turn called the latter *ab-mūl.wa* (a deaf person), implying that only the practiced ear of an *àryô.to* can judge the distance of a turtle so correctly so as to be able to harpoon it in pitch darkness, or *gū.gma-tòng* (leaf of the *Trigonostemon longifolius*) 'in allusion to the practice, current among the inland tribes, of using these leaves for the cure of fever, but to which *àryô.to* rarely have recourse, as they believe the scent prevents turtles from approaching a canoe in which there are any persons who have recently employed this remedy' (p. 35).

As the mainstay of these groups, the pig and the turtle were for the *ē.remtâ.ga* and *àryô.to* respectively, recurring elements within ritual and taboo. Beginning with the onset of puberty, Andamanese youth traditionally underwent an *â.kà-yâ.p*, a period of avoidance of key foods via a series of long duration fasts (*gū.mul*) that were meant to test their strength of endurance and powers of self-denial. For girls, this phase began with the first menstruation (this period also signified the termination of her childhood name and the permanent adoption of her flower name, that of a tree which was flowering during that time). For boys, this was decided by friends and relatives. The idea was that from that point onwards, the youth would abstain from all the chief foods of the community, but the fasts would be undertaken one at a time so as to avoid starvation.

The order and nature of these foods varied according to the community. For instance, among the coast dwellers, the initial taboos began with key foods such as turtle, dugong, porpoise, and certain species of fish and shell-fish, whereas among the forest dwellers, wild pig, honey, the larvae of the Great Capricornis beetle (*oiyum*) and certain species of fish from the creeks (e.g. *nyuri*) were taboo. A number of other species were also abstained from during the period of adolescence – monitor lizard, the flying fox (*Pteropus* sp.), certain species of fish (e.g. *komar*), bird species, varieties of mangrove seeds, edible grubs (*pata* and *cokele*), several species of edible roots, and several vegetable foods.

In some groups, those who had not undergone these initiation ceremonies were prohibited from eating turtles, dugongs, etc. (Radcliffe-Brown 1922). Each major fast was brought to an end by a feast solely involving the species which was previously forbidden. Several other dietary prohibitions relating to frequently utilized species were also in place as part of personal milestones (e.g. pregnancies) and during events such as births, deaths, and mourning periods.

Those who disregarded prohibitions were likely to incur the wrath of the supreme being *Puluga* or associated supernatural beings. *Puluga* has been proposed as a figure associated with the 'thunder-complex', a distinct ethnographic marker common to the Andamanese, as well as the Orang Asli of the Malay Peninsula and the Philippine Agta (Aeta) (Blust 1981, 1991, 2013). Though the exact nature and gender of this being – also referred to as the wind *Puluga-Biliku* (*Dare* among the Ongee) – is difficult to ascertain from colonial accounts, disregard for taboo (especially those relating to mockery and inappropriate treatment of animals, plants, and hunted meat), and other shared taboos such as the mistreatment of cicadas and burning of bees-wax (used to bind rope on harpoons) was believed to result in retribution via punitive thunderstorms.

Man (1883) reports that the wood of the *al.aba-* tree – the source of fibre for making harpoon lines and turtle nets – was not to be used for cooking turtle, this act being considered abhorrent to *mai.a. ō.gar-* (the moon). Similarly, there were a number of taboos associated with plants that played a role in hunting: *Hibiscus tiliaceus, Myristica longifolia, Anadendron paniculatum, Ficus laccifera, Gnetum edule* – all used in turtle hunting, and *Tetranthera lancaefolia* from which pig arrows were made, were associated with prohibitions.

During the initial part of the rainy season, the palm, *Caryota sobolifera, Dioscorea* yams, and edible roots were treated as having been sequestered by *Puluga* for their own consumption. While allowances were made for the consumption of fallen seeds (i.e. those found on the ground from the previous year) of *Entada pursaetha*, plucking the seeds was likely to invite retribution from *Puluga* in the form of a deluge or thunderstorm.

According to Radcliffe-Brown (1922), civet cat (*Paradoxurus*) and monitor lizard were allowed to be eaten largely in the rainy season. Towards the end of the rainy season – in the short period of unsettled weather – two delicacies, the larvae of the Great Capricornis beetle (*oiyum*) and that of the cicada (*butu*, also considered the offspring of Biliku (*Biliku ot-tire*)) were permitted for consumption. However, in other seasons, the *butu*, which are 'seldom seen, and cannot be disturbed' were connected with a strong taboo, whereby no noisy work was undertaken in the morning and evening when the cicada 'sing' because this would cause the *butu*'s head to ache and would in turn bring bad weather. Similarly, digging up yams was usually accompanied by the practice of removing the lower portion and replacing the crown of the yam attached to the climber, in the hope of deceiving *Puluga* who was the owner of yams. This practice also likely served in their propagation.

Hunted pig too was carved (a badly quartered animal aroused *Puluga*'s temper) and prepared the right way (by boiling as opposed to roasting) so as to avoid detection by *Puluga* and the *chôl-* (*chôl-* is another class of malevolent spirits who were *Puluga*'s associates. *chôl-* is also the local name of the clever racket-tailed drongo according to Radcliffe-Brown) who would punish the delinquents.

In North Andaman, Radcliffe-Brown (1922) described an association between the souls of unborn babies and little children who died before they were weaned, and the green pigeon and the *Ficus laccifera* which the pigeon frequently feeds on. The souls were believed to reside in these trees as a consequence of which these trees were protected, with the exception of the removal of bark from the aerial roots for personal ornaments and fibre for bow strings.

### *Little Andaman*

Recently, long-term work among the Ongee of Little Andaman (*Gaubolambe* to the Ongee), by Pandya (1993, 2009), offers a detailed ethnography that also points to the centrality of pigs, turtles, and dugongs. The Ongee too are primarily divided into pig hunters (*Eahansekwe*) and turtle hunters

(*Eahmbelakwe*) (Radcliffe-Brown 1922; Man 1883, Pandya 1993, 2009) within the four major territorial clans.

Similar to the Great Andamanese categorizations, turtle hunters are known as *gebogealh* (those who see well), and pig hunters as *gekalegebaro* (those who hear well). Ongee marriages are usually between pig hunters and turtle hunters across territorially identified and proscribed clans. Each clan comes under the oversight of a mythical bird appointed by the creators of the Ongee to inform the spirits that the Ongee were moving in the prescribed manner, according to the seasonal cycle (*monatandunamey*) of the winds (*Dare – Biliku* of the Great Andamanese, *Mayakangne?, Kwalakangne?*), and the spirts (*tomya*), that ensured that hunting and gathering were adequately distributed within the shared spaces (Pandya 1993, 2009).

According to Pandya (1993, p. 70), 'the spirits, like the Ongees come to the island to hunt and gather and take the limited food resources of the island, thus making food either available or unavailable. It is because of the food-seeking spirits who visit the island that the Ongees say, '*All food is not to be found in the same season and in all places*'. Food, therefore, is subject to seasonal variations for the Ongee hunters and gatherers. When the spirits are out at sea hunting and collecting food for themselves, the humans are expected to be in the forest, hunting, and collecting their own food. The food consumed by the spirits in a given season is proscribed for the Ongees, who cannot gather or hunt, or even consume these items during that period. Consequently, the Ongee community undertakes seasonal translocations'.

Therefore, in the monsoon months from May to September, when the spirits hunting in the sea bring about violent weather that make it impossible to hunt in the sea, Ongee families move into the encampments of the pig-hunters in the forest who supply them with hunted pig. They speak of this as a time when '*Spirits are hunting turtles, making the sea a rough place and we humans are in the forest surrounded by ripe forest fruits that pigs feed on and they are fat and heavy! They are ready to be heard and taken (that is, ready to be hunted)*' (Pandya 2009, p. 249). As the weather settles between October and February, the pig hunters in turn come to the coast to stay with the turtle hunters who hunt turtle (and in rare cases, dugong in late October). During these months, there is a prohibition on hunting pigs as the spirits are hunting them.

Only novices, for whom transgression is a critical part of their initiation (*tanageru*), are allowed to go into this spirit world and hunt pig. The successful culmination of the *tanageru* ritual indicates that the initiate has demonstrated the qualities of a successful hunter. It earns him the right to hunt in the forests and the sea, as well as to transmit the knowledge and techniques related not only to hunting but also to the movement of the spirits, seasons, and winds.

During March and April – which is the dry phase (*Torale*) – the Ongee expel the spirits from the island through a series of 'offensive' rituals that include the consumption of honey (the favourite food of the spirits). This is believed to make the spirits angry and induce them to come down again bringing with them wet weather, during which period they get trapped in fruits and honeycombs, which in turn are consumed by the women who fall pregnant as a result.

As mentioned before, dugong hunts are typically rare events usually undertaken in late October. In Ongee myth, a large body of narratives known as the *jugey* describes the creation of the dugong – a creature that is simultaneously pig and turtle – born out of the conflict between two powerful beings, the monitor lizard and the civet cat who enticed the pigs from the forest out to the sea, while inviting the turtles to take the place of the pigs. This resulted in the creation of a creature with the partial behaviours and morphologies of both its ancestors that is unable to live either in deep water or land. Every full moon, singing sessions are carried out by the Ongee, followed by a ritual hunt in the morning with an aim of mercy killing or putting the dugong out of its misery (Pandya 2009).

The Ongee (as well as other Andamanese communities) have been known to preserve the skulls of hunted pigs, turtles, and dugongs – animals that are central to their existence. Once the animals are cooked, the skulls are preserved carefully and kept in Ongee homes with the intention of signifying to living animals (through smell) the presence of these animals. This ensures that they remain in place without leaving their hunting grounds, guaranteeing success in later hunts. Pandya (2009) also speaks of a similar practice among the Jarawa during pig hunts where fistfuls of pig blood are thrown in the air for its smell, in order to deceive other pigs and spirits into believing that no pigs have been harmed.

According to Ongee myth, the Ongees, who were fond of eating the birds, were put off from this by the birds (firstly, the Andaman scarlet minivet and then the white-breasted kingfisher) that were instrumental in stealing fire from the spirits to give to the Ongee to win their favour. Four birds (*Chouling, Gaye?, Amiya* and *Amie*), one for each clan, decided to stay with the Ongees and guide them about spirits and body designs. Designs on the body correspond to species found in the season in which an individual was born (*Barey?* – turtle, *Alakaye* – dugong, *Enetandabokatale* – fruit of *daboja* – *Bruguiera gymnorhiza*).

Birds and animals also signal the end of the dry season (*Torale*) in May and the arrival of the monsoon (*nakwarbe* ushered in by *Mayakangne?*). The first signal is the disappearance of birds such as *bele-bele* and *kentamale* (drongos, shrikes, swallows, and martins), considered friends of the Ongee who guide them to the honeycombs. The second series of signs are read in cloud formations that signify the arrival of rain that flood the honeycombs in the forest. The third and the most definitive sign is the presence of furrows and uprooted soil in the forest that the Ongee refer to as the pig's digging work. This signifies the end of the jackfruit season, when the Ongee say that since all the fallen jackfruit have been collected by the women, the pigs, which have now become fat and heavy are frantically uprooting the soil for *gegi* tubers to feed their piglets.

Pandya (1993) also recounts Ongee myths in which the spirits are angry because the women and children collected too many turtle eggs on the island of Tetale – North Brother, a small island north of Little Andaman. As a result, the spirits decided that only stones would be left there, and the Ongee learnt from the spirits that they were not to have eggs and turtle during that season. Apart from the turtle and the dugong, the nautilus shell is of special importance to the community. In addition to its quotidian use as a drinking container, the Ongee carry the nautilus shell on all canoe journeys as they believe it is endowed with unique capacities. In Ongee myth, young boys were sent in nautilus shells to meet the spirits for negotiation and re-establish the availability of resources in a time of conflict. During the *tanageru* ritual, the novices themselves are referred to as a *naratakwange*, a chambered nautilus shell.

## Lessons and challenges for conservation

Explorations of ethnographies of Andamanese communities reveal the existence of myths, rituals, and taboos. Each of the broad categories of RHTs outlined by Colding and Folke (2001) can be found within islander cosmologies, that is segment taboos, temporal taboos, life-history taboos, species-specific taboos, habitat taboos, and method taboos. As is the case among hunter-gatherers in other places as well, these are of an overlapping nature. In general, myths and stories not only bring into the psyche of the hunter-gatherer a great deal of familiarity with a large number of species but also provide rules concerning use of resources that are rare, highly prized, or are integral to the survival of communities.

As pointed out by Radcliffe-Brown (1922), the exploitation of high value foods such as honey and yams – foods containing sugar and starch that were on the whole rare and highly prized – was guided by cultural injunctions in which the anger of the spirits, manifested in thunderstorms and other forms of adverse weather, has consequences for the availability of food and other resources. Translocationary lifestyles controlled by community-level rules are particularly important for small bands occupying island systems which are typically depauperate in mammalian fauna. Pigs and other terrestrial resources such as yams and honey need to be managed to avoid overexploitation.

For the Ongee and other Andamanese communities, encoded within stories, myths, and rituals are rules and prohibitions – specific times and seasons to hunt pigs, turtles, and dugong and designated places and territorial boundaries for hunting – that ensure resource availability over the long term. Multigenerational knowledge transfer, as well as involvement of the whole community, is key to continued sustainability. Unlike modern exercises in conservation and sustainable use that are based on scientific measurement (e.g. hunting quotas), sustainability among foragers continues to be ensured through everyday practice and associated cultural engagements that are inseparable from each other. Conservation in this format is locally relevant, largely voluntary, organic and self-devised, and self-imposed by the community, rendering them less costly and less divisive than formal top-down interventions imposed and enforced by the state.

As pointed out by Colding and Folke (2001), such adaptive responses are characteristic and work best in systems with effective community property rights, where resources are held by a discrete set of users who also have the wherewithal to regulate their use by excluding other groups. Historically, these islands served as ideal systems for such forms of governance, whereas in recent years, settler dynamics and state intervention has posed problems at several levels. Conservation by the state has resulted in a set of imposed norms and restrictions on movement of many groups, compounded by infrastructural development and the influx of outsiders. Recent settler incursions into the hunting landscapes of the Ongee and Jarawa have impacts on social and nutritional dynamics, especially along linear intrusions such as the Andaman Trunk Road (e.g. see Sekhsaria and Pandya 2010).

The informal RHT systems described earlier do not serve conservation alone. In fact, a key beneficial feature is their holistic treatment of several societal axes, in that a variety of linked components are dealt with together – for example, rules restricting overexploitation and population growth. Unlike conventional conservation projects that often attempt standalone interventions, these forms of adaptive responses organically take care of interlinkages. At the same time, it is very important to note that in the context of this chapter, myths and symbols are isolated in order to speculate on potential ecological functionality of ritual in relation to island resources, at least in the case of terrestrial ones that are somewhat finite.

However, in local cosmologies these are not at all discrete. Instead, they are tied up strongly in complex cosmologies which are holistic and serve a variety of functions other than just ecological sustainability and food availability. For instance, some have bearing on social status. These therefore need to be viewed as interrelated components of a single, yet complex system or cosmology. Long-term ethnographies that attempt such an understanding are already available through the work of anthropologists such as Visvajit Pandya (1993, 2009).

Given present day pressures and threats, a recurring concern has been whether the persistence of these systems thus far has been a result of happy coincidences (e.g. isolation and permissive environmental conditions) or are eventually doomed to extinction (Tuck-Po 2013). Recent anthropological scholarship on other negrito communities in Southeast Asia, however, points to the extreme resilience of hunter-gatherer communities, on account of a range of strategies such as mobility, adaptability, exchange relationships, and superlative strategies in dealing with aspects of risk and uncertainty (Bellwood 2007, Chan 2007, Tuck-Po 2013) in a way that epitomises Folke's (2006) definition of resilience.

At the local scale, there are adaptations to change. Among the Ongee, the *tanageru* initiation ritual is a key personal milestone and a necessary one for the community. However, keeping in view the dwindling population, the shrinking forest, and the declining numbers of pigs as a consequence of hunting by outsiders, annual initiation rituals have been carried out over two years instead of one.

However, at a larger scale, there are greater uncertainties. Interwoven within their own myths are recent experiences of change that are more or less insurmountable. As pointed out by Pandya (2009), analysing their own predicament in the context of contact with outsiders, the Ongee themselves identify their situation with that of the dugong, as being unable to cope with change. Although a limited amount of contact and exchange facilitated the acquisition of iron and a few other necessities, hostility to outsiders was an important feature of Andamanese groups which enabled them to hold their own for several centuries.

Colonial inroads into the islands eventually changed these conditions. Once established, efforts were undertaken by the colonial administration to induce translocationary groups to settle. Dogs were imported onto the islands by the British and given to different communities to improve their hunts and to restrict seasonal translocations. Cipriani (1966) is of the opinion that while the introduction of dogs into the archipelago resulted in an increase in the short-term availability of meat, since dogs aided hunting, the long-term impact has been negative. An explosive growth of the dog population resulted in frequent hunting of young pigs by packs of dogs. Cooper (1990) points out that although from a historical perspective, the introduction of dogs was a minor event, this led to a reduction in the numbers of pigs and the size of their kitchen middens. For the islanders, this ushered in an era of immense change. For the islanders of North Andaman, the period before colonization came to be imprinted in their psyche as '*bibipoiye*' (days when there were no dogs).

## Conclusion

The existence of local systems of regulation does not ensure that all forms of traditional use systems are always benign or just. There is a tendency for valorisation of traditional systems painting them exclusively as benign (e.g. Shiva 1988). There are also arguments that allude to special gender relationships (e.g. the ecofeminist argument by Shiva (1988) that paint women as nurturers), romanticized traditional ethics of indigenous people (Pereira and Seabrook 1990, Pereira 1992), etc.

This has been labelled by Sinha et al. (1997) as the 'new traditionalist discourse', cautioning against the simplistic viewing of traditionalism as benign from the perspective of social justice. However, among Andamanese communities it is worth noting that while perhaps there are more restrictions for women, it is unlikely that there are major impacts on nutrition and nourishment. While women may not be as active in hunting as men, they acquire the role of primary decision makers in a variety of associated spheres.

Sustainable use in island systems such as the Andamans can be strengthened by ensuring autonomy and status quo in existing indigenous ways of life, and also by buffering such groups from outside influences to a certain extent. The understanding of resource use as an adaptive process interlinked within local cosmologies is often lacking in standalone state interventions, as is the requirement for adequate space for small bands of people. Furthermore, sustainable use in the islands will also benefit from parallel interventions with local settler communities in designing more compatible ways of sharing resources and space.

## References

Bellwood, P. (2007). *Prehistory of the Indo-Malaysian archipelago*. Canberra: ANU Press. http://doi.org/10.22459/PIMA.03.2007

Blust, R. (1981). Linguistic evidence for some early Austronesian taboos. *American Anthropologist*, *83*(2), 285–319. https://doi.org/10.1525/aa.1981.83.2.02a00020

Blust, R. (1991). On the limits of the "thunder complex" in Australasia. *Anthropos*, *86*, 517–528.

Blust, R. (2013). Terror from the sky: Unconventional linguistic clues to the Negrito past. *Human Biology*, *85*(1–3), 401–416. DOI: 10.3378/027.085.0319

Chan, H. (2007). *Survival in the rainforest: Change and resilience among the Punan Vuhang of eastern Sarawak, Malaysia*. [Doctoral dissertation, University of Helsinki]. Helda Helsinki.

Cipriani, L. (1966). *The Andaman islanders*. London: Weidenfeld & Nicholson.

Colding, J., & Folke, C. (2001). Social taboos: 'Invisible' systems of local resource management and biological conservation. *Ecological Applications*, *11*(2), 584–600. https://doi.org/10.1890/1051-0761(2001)011[0584:STISOL]2.0.CO;2

Colebrooke, R. H. (1799). On the Andaman Islands. *Asiatic Researches*, *XXVII*, 401–415.

Cooper, Z. (1990). The end of '*Bibipoiye*' (dog not) days in the Andamans. In B. Meehan & N. White (Eds.), *Hunter-Gatherer Demography: Past and present* (pp. 117–125). Sydney: University of Sydney.

Dickman, A., Johnson, P. J., van Kesteren, F., & Macdonald, D. W. (2015). The moral basis for conservation: How is it affected by culture? *Frontiers in Ecology and the Environment*, *13*(6), 325–331. https://doi.org/10.1890/140056

Folke, C. (2006). Resilience: The emergence of a perspective for social-ecological systems analyses. *Global Environmental Change*, *16*(3), 253–267. DOI:10.1016/j.gloenvcha.2006.04.002

Gadgil, M., & Guha, R. (1993). *This fissured land: An ecological history of India*. Oxford: Oxford University Press. DOI:10.1093/acprof:oso/9780198077442.001.0001

Harris, M. (1971). *Culture, man, and nature: An introduction to general anthropology*. New York: Thomas Y. Crowell Company. https://doi.org/10.1002/ajpa.1330370311

Johannes, R. (1981). *Words of the lagoon: Fishing and marine lore in the Palau district of Micronesia*. Berkeley, CA: University of California Press.

Johannes, R. E. (1978). Traditional marine conservation methods in Oceania and their demise. *Annual Review of Ecology and Systematics*, *9*, 349–364. https://doi.org/10.1146/annurev.es.09.110178.002025

Man, E.H. (1883). On the aboriginal inhabitants of the Andaman Islands. *The Journal of the Anthropological Institute of Great Britain and Ireland*, *12*, 69–175. https://doi.org/10.2307/2841843, https://doi.org/10.2307/2841953

Mouat, F. J. (1863). *Adventures and researches among the Andaman Islanders*. London: Hurst and Blackett.

Pandya, V. (1993). *Above the forest: A study of Andamanese ethnoanemology, cosmology, and the power of ritual*. Oxford: Oxford University Press.

Pandya, V. (2009). *In the forest: Visual and material worlds of Andamanese history (1858–2006)*. Lanham, MD: University Press of America. DOI:10.1017/S0021911810002792

Pereira, W. (1992). The sustainable lifestyle of the Warlis. *India International Centre Quarterly*, *19*(1–2), 188–204. www.jstor.org/stable/23002229

Pereira, W., & Seabrook, J. (1990). *Asking the earth: The spread of unsustainable development*. Mapusa: The Other India Press.

Portman, M.V. (1899). *A history of our relations with the Andamanese* (Vols. 1 and 2). Kolkata: Office of the Superintendent of Government Printing.

Radcliffe-Brown, A. R. (1922). *The Andaman Islanders: A study in social anthropology*. London: Cambridge University Press.

Rappaport, R. A. (1968). *Pigs for the ancestors: Ritual in the ecology of a New Guinea people*. New Haven, CT: Yale University Press.

Reichel-Dolmatoff, G. (1976). Cosmology as ecological analysis: A view from the rainforest. *Man*, *11*(3), 307–318.

Sekhsaria, P., & Pandya, V. (Eds.). (2010). *The Jarawa tribal reserve dossier: Cultural and biological diversities in the Andaman Islands*. London: UNESCO.

Shiva, V. (1988). *Staying alive: Women, ecology and survival in India*. New Delhi: Kali for Women.

Sinha, S., Gururani, S., & Greenberg, B. (1997). The 'new traditionalist' discourse of Indian environmentalism. *The Journal of Peasant Studies*, *24*(3), 65–99. https://doi.org/10.1080/03066159708438643

Tuck-Po, L. (2013). Making friends in the rainforest: "Negrito" adaptation to risk and uncertainty. *Human Biology*, *85*(1–3), 417–444. DOI:10.3378/027.085.0320

# 8 Rethinking indigenous hunting in Northeastern India

## Some lessons for academics and practitioners

*Ambika Aiyadurai and Sayan Banerjee*

### Introduction

'Which animal did Salman Khan hunt?' asked a Mjiu Mishmi villager in Anjaw district of Arunachal Pradesh in 2005. His voice was filled with curiosity. The mammal and bird field guides that we carried came in handy for starting a dialogue and breaking the ice with locals in India's northeastern state. Pictures of the blackbuck *(Antilope cervicapra)* were new to most of them. Suddenly, villagers surrounded the book, jostling with each other to take a look at these animals. Many people in India first heard about the blackbuck from the media coverage about a poaching case by a high-profile Bollywood actor, which naturally made prime-time news.

The Mishmi villagers giggled after seeing the picture of 'Salman Khan *wala jaanwar* (in English: Salman Khan's animal)'. One of the villagers explained, 'He hunted these animals, and he had to go to prison. But, here we hunt animals every day'. The news of the arrest had reached this remote area through radio, but the villagers could not figure out what all the fuss was about. The reactions of the villagers highlight two critical facets of wildlife hunting in north-east India. First, hunting is commonly practised in large parts of the region, and second, there is a low level of awareness of India's wildlife laws. Hunting has always been part of the culture of local people, and this practice continues even today, at least in the villages located away from the state capitals.

People here take pride in hunting, and their hunting exploits are often talked about openly and freely (See Figure 8.1). For example, the recent news of the killing of a king cobra in Arunachal Pradesh and hunting in Nagaland during the COVID-19 lockdown (April 2020) went viral on social media.

Wildlife hunting in north-east India has, until recently, been studied as an ecological 'problem'. Increasingly though, scholars are examining its social and human dimensions. Sociological approaches reveal hunting as a complex, multifaceted, and culturally embedded practice (Aiyadurai 2011). In this chapter, we start with an overview of studies in north-east India on hunting, followed by local indigenous views on hunting. The chapter will discuss crucial challenges for conservation that drastically differ from

DOI: 10.4324/9781003343493-11

*Figure 8.1* Skull display in Nagaland
*Photo: Sayan Banerjee*

mainland India alongside addressing challenges for academic and policy engagement with wildlife hunting and sustainable use.

## Studies on wildlife hunting

North-east India, as a part of two 'biodiversity hotspots' namely Eastern Himalaya and Indo-Malaya, (Myers et al. 2000) has been subject to a growing interest among wildlife biologists. Several studies, surveys, and anecdotal records from north-east India suggest that hunting is a social–cultural practice, and with some finding it a significant threat to many species (Datta et al. 2008; Aiyadurai et al. 2010; Velho et al. 2012). For instance, a study in Namdapha National Park in Arunachal Pradesh indicated the absence of large carnivores and prey base and reported an 'empty forest' landscape (Datta et al. 2008).

Studies on hunting, and related activities, typically include preliminary information on species profiles, distribution, hunting patterns, intensity and drivers, off-take rates, and market demands (Hilaluddin and Ghose 2005; Tynsong et al. 2012; Velho and Laurance 2012; Selvan et al. 2013; Bhupathy et al. 2013; Singh et al. 2014; Naro et al. 2015; D' Cruz et al. 2018). Aiyadurai et al. (2010) reported 33 species of mammals hunted in Arunachal Pradesh and analysed the reasons for hunting, highlighting the social and cultural linkages. Hilaluddin and Ghose (2005) listed the hunting of 134 wild animals (mammals, birds, and reptiles) from Nagaland, Mizoram, and Arunachal Pradesh. Chutia (2010) documented 43 species of mammals hunted in Nagaland and Arunachal Pradesh. While these studies provide valuable insights from a region that is mostly understudied, conclusions can hardly be drawn from these studies on the sustainability of hunting practices.

## Indigenous view of wildlife hunting

In 1972, the *Wildlife (Protection) Act* made hunting a criminal offence. This Act, along with other similar protectionist legislation was detrimental to forest-dependent communities, the majority of whom were scheduled tribes and marginalised dalit groups, as their traditional livelihoods were deemed illegal overnight. Though for many, hunting is an activity deeply embedded in culture and livelihood strategies, making such a law ineffective in tribal-dominated regions. With little cognizance of the social–cultural realities of hunting in north-east India, the *Wildlife (Protection) Act, 1972* has failed in this region (Datta 2007, Aiyadurai and Velho 2018).

One of the reasons attributed to the prevalence of hunting in northeastern India is the stronger ownership of communities over their land and forests in comparison with the governmental agencies in the region (Datta Roy 2018). Moreover, hunting is not only significantly linked to local livelihoods and food security but also it is accepted as a social practice that expresses tradition and identity. Therefore, hunting is both culturally significant and locally accepted.

For indigenous people, the practice of hunting goes beyond simply killing an animal for food to encompass a more extensive, complex, and dense network of exchange between humans, animals, and the spirit world. For example, the role of *uyu* (spirit) in defining the relationship between the hunter and the animal is crucial among the Nyshis (Aisher 2007).

Similarly, among the Idu Mishmi, the *khə̄nyū* spirits play a significant role during hunting, farming, and fishing. *Ŋgōlō*, the deity of the higher altitudes, is said to be the 'caretaker' of the wild animals; the animals must, therefore, not be killed without certain obligations (Aiyadurai 2018, Nijhawan and Mihu 2020). This concept of forbidden species known as *mísū* is central, and people follow codes of conduct and behaviour so that they receive blessings and success in farming, hunting, and other resource activities (Blench 2016). If the people fail to satisfy the spirits, harvests may fail, hunts can be unsuccessful, and some unfortunate events may occur to a person within a family or in the village. Such practices are found across several indigenous groups in northeastern India (Janaki et al. 2020), where local communities follow animistic traditions in which the role of shamans is crucial in many practices, including farming, healing, and hunting (See Figure 8.2).

The significance of wild meat is not just restricted to protein intake, but rather meat becomes part of a more extensive network to strengthen social ties and kinship. For example, wild meat is offered as a form of bride price during weddings and is regarded as a status symbol by the local communities in Anjaw district. Among the Adis, men offer orange-bellied squirrels (*Dremomys lokriah*) to the bride's family during

*Figure 8.2* Idu Mishmi Shaman (Igu)

*Photo: Ambika Aiyadurai*

*Figure 8.3* Otter skin
*Photo: Ambika Aiyadurai*

weddings (Das and Shukla 2007). Wild meat or associated products (skulls, skins, or horns) are gifted to visiting government officers as a goodwill gesture. Many of these items become decorative showpieces in the drawing rooms of government employees and other elite members who reside in the towns (See Figure 8.3).

Delving deeper into the worldview of indigenous people of the region provides insight into alternate ways of understanding human–nature relations. The social worlds of many of the indigenous peoples in north-east India include humans and non-humans (animals, plants, landscapes, spirits). The relations of humans with animals are imbued with cosmological meaning and pragmatism. In their philosophies, the lives of humans and non-humans overlap, thus preventing the dominance of a human-centric society. Therefore, people consider themselves a part of a broader ecological system fostering kin relations, which can extend to assigning personhood to non-human beings.

Scholars have highlighted that the relations with the natural world cannot be understood as monolithic and simplistic, but that social attitudes to animals are complex, diverse, and multifaceted (Knight 2004, Morris 1998). The interactional encounters between humans and non-humans demonstrate a complex multispecies inter-subjectivity in which all human lives and cultures are enmeshed. The multiple local or place-based understandings of nature are contradictory to the way modern societies see nature just as a 'resource' to be exploited for human utility. Indigenous worldviews may thus have lessons for navigating the global environmental crises that are upon us (Aisher and Damodaran 2016).

## Conservation initiatives

The wide-spread prevalence of hunting in north-eastern India has led to a number of conservation interventions to tackle hunting. Some projects call locals to take a stand against hunting (Datta 2007). 'Ex-hunters' are hired as field assistants, trackers, or patrol staff for wildlife monitoring programmes implemented by NGOs in collaboration with forest departments.

In 2003, the Wildlife Trust of India and the Arunachal Pradesh Forest Department jointly undertook a conservation initiative for hornbills by introducing artificial beaks made of fibreglass or wood to replace original hornbill beaks in order to prevent Nyishis from hunting hornbills (See Figure 8.4). For the Nyishi men, the hornbill beak is a symbol of manhood, valour, and identity and hunting a hornbill is a rite of passage. The casque, beak, and feathers are used as a part of a traditional headgear – *pudum*. Wearing of the headgear is mandatory for Nyishi men, especially during festivals, marriages, and social gatherings.

Following the introduction of the artificial beaks, the Village Development Councils announced a fine of Rs. 5,000 for any person found hunting hornbills. This initiative was an acclaimed success for effective community

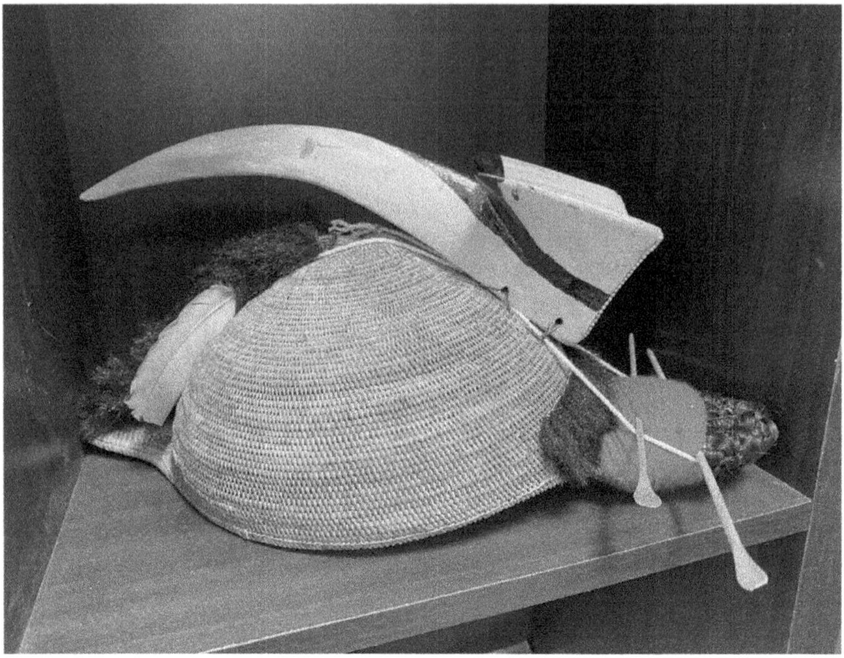

*Figure 8.4* Head gear with artificial hornbill beak
*Photo: Ambika Aiyadurai*

engagement to reduce hunting. Kumar and Riba (2015) showed that around 60% of the 71 Nyishi people interviewed preferred headgear made with the artificial beak, while 33% preferred headgear with the original beak and the rest preferred both.

In 2011, a citizen-funded hornbill nest adoption programme was initiated by the Nature Conservation Foundation in Pakke Tiger Reserve in Arunachal Pradesh, where local Nyishi people were employed as hornbill nest watchers (Rane and Datta 2015). The project reported increased nesting and reproductive success of hornbills. In addition, vigilance by nest watchers has resulted in reduced logging and hunting in the tiger reserve and community ownership of resources (NCF-India, n.d.; Rane and Datta 2015; Rahman 2016). Several other conservation projects such as community-based bird tourism in Eaglenest Wildlife Sanctuary in Arunachal Pradesh, Amur Falcon (*Falco amurensis*) conservation in Nagaland, and Greater Adjutant stork (*Leptoptilos dubius*) conservation in Assam have gained popularity for their significant reduction in hunting, habitat and species protection, and employment opportunities for alternate livelihoods around community-centric activities.

Yet another conservation approach that has gained currency in the region is the establishment of community conserved areas (CCAs). These are espoused as an effective way to conserve forests and wildlife and are primarily managed by the local community with the help of governmental and non-governmental organisations. Although autonomous community efforts to conserve forests have been present for a long time, in 1998, the Khonoma village in Nagaland first institutionalised this CCA model by establishing the Khonoma Nature Conservation and Tragopan Sanctuary for the conservation of Nagaland's state bird, the Blyth's Tragopan (*Tragopan blythii*), as well as other important species. Soon, with the help of NGOs and a willing Forest Department, more such CCAs were established in Nagaland. In 2015, 407 CCAs of various sizes were documented in Nagaland (TERI 2015). CCAs have been set up in Arunachal Pradesh as well, and these are considerably larger in size, with CCAs in West Kameng and Tawang spanning more than 100 sq km each (Krishnan et al. 2012).

However, little is known about the social impacts of such interventions, an issue that calls for more in-depth assessments. A closer examination of these projects reveals new tensions within local communities, especially in relation to decision making and benefit sharing from the new livelihood opportunities that have emerged. A shift from intrinsic conservation, based on the cultural value of a species, to one that relies on its commercial value or its ability to meet developmental aspirations was also observed (Aiyadurai and Banerjee 2019).

## Outcomes of conservation initiatives

Across many parts of the northeast region, there is a shift towards species-focused conservation and sustainable use measures, though these are not without challenges. For instance, in Nagaland the community constitutionally owns the land and thus, national wildlife laws have little acceptance and enforceability in this area (Banerjee 2016). There is resistance to attempts to enforce national legislation on wildlife conservation (Aiyadurai 2018).

In many areas, people think that such 'Indian' laws are against their tradition and feel that these laws are to blame for the long-term injustices they have suffered. This tussle between national and customary laws has not been adequately addressed in wildlife conservation discourse. Even local hunting regulations that villagers themselves have formulated are not followed by everyone. Hunting in such a situation is like an 'open secret'. Another challenge is the geographical spread of the forests that may cut across multiple village boundaries. Such regulations only succeed if all villages surrounding the forest follow the rules. Even in places where such customary laws work, people still go hunting due to their preference for wild meat over domestic meat.

There are places where the amount and variety of wild meat consumption may have decreased, but the preference for wild meat remains. In Nagaland, high-income families see wild meat as a luxury (Hilaluddin and Ghose

2005). As elsewhere in the region, people living in comparatively remote areas have limited access to markets and are mostly dependent on wild meat. Those who have migrated to cities and towns still prefer the taste of wild meat. Locals believe that wild meat is 'purer' than meat of domesticated animals and as a result, people may pay up to five times more for wild meat than for the meat of domestic animals (Hilaluddin and Ghose 2005). As a consequence, wild meat is widely available in town markets. On the other hand, due to people's allegiance towards village councils and other local institutions, regulations against hunting have been successful in some villages.

CCAs face enormous challenges due to their relative invisibility to policymakers, the absence of mainstream legal support, disinterest from youth, and a range of social costs (Krishnan et al. 2012). While the creation of CCAs in north-east India has received significant attention among conservation practitioners, rigorous studies on their ecological and social impacts have been scarce. It is difficult to estimate if these regulations have caused an ethical shift about eating wild meat among people. Sound research and engagement are necessary to ascertain the sustainable use of wildlife within the interplay of politics, legality, lifestyle choices, livelihoods, and ecology.

Newer conservation initiatives in the region have the potential to develop serious engagements with the community, increase landscape protection, reduce hunting and prevent unsustainable resource harvest practices, thereby ensuring long-term ecosystem benefits at lower costs (Kothari et al. 2000; Pathak 2009). However, there is considerable uncertainty around the sustainability of hunting in this region that requires more investigation. For instance, how do people conceptualise the sustainable use of wildlife? How should local, regional, and global laws be streamlined so that the combined interests of local communities and wildlife conservation are addressed? How do traditional local institutions impact people's behaviour towards wildlife and how should such institutions be strengthened?

Action-based research is needed that can document and further establish protocols of sustainable forest dependency. Sethy and Chauhan (2012) in their study on the conservation of the Malayan sun bear (*Helarctus malayanus*) in Nagaland, support the enforcement of wildlife laws and reduced forest dependency in order to sustain the population of this species. However, measures such as these may fail due to their disengagement with the socio-political and historical realities of traditional societies (Aiyadurai 2011). Prior to suggesting local conservation solutions, perhaps conservationists and academics could explore how locals perceive habitat loss and species decline, and the meanings they assign to their forests and wildlife?

For instance, Idu Mishmis claim that the tigers in their landscape are surviving because of the social taboos they observe, which in turn is helping to conserve tigers and other wildlife (Aiyadurai 2016, Nijhawan 2018, Nijhawan and Mihu 2020). Mishmis assert that they too are conservationists and ask for their role in conservation to be acknowledged. Conservation narratives are largely tied up with wildlife census, discoveries of species, recognition by global conservation communities, and funding agencies – all of which tends to valorise the contribution of wildlife biologists. Local communities become mere statistics, with the occasional scholar romanticising their 'exotic' lifestyles.

As Ulrich Beck (2010) notes, 'if "the environment" only includes everything which is not human, not social, then the concept is sociologically empty. If the concept includes human action and society, then it is scientifically mistaken and politically suicidal'. He further observes that the discourse on environment politics is an 'expert and elitist discourse', and one where ordinary peoples' views are not counted. This chapter echoes the same sentiments. The voices of wildlife enthusiasts and experts are heard loudly and clearly, through academic and popular publications, social media, and wildlife documentaries that shrink the space for the perceptions of locals.

Among wildlife biologists, there has been great reluctance to take onboard local expertise, which is routinely disregarded or treated as anecdotal. Researchers often do not consider issues outside the purview of wildlife research, particularly when they are fixated on one particular species or habitat. Studies on hunting from north-east India often contain considerable information about the species and less, or sometimes nothing at all, about the social dimensions of such hunting. Where local communities are mentioned, they are often implicitly framed as 'poachers' or 'ecological criminals'.

Changing this will require conservation researchers to take note of local social realities and break out of their disciplinary straitjackets. Efforts to find new ways of engaging with local communities and paying attention to community insights about wildlife are few. There are also severe pedagogical limitations that produce this outcome. Wildlife researchers carrying out studies or surveys on hunting are often not trained in sociological concepts and research methods. Often times, they are simply unaware of them. As a result, stories from the ground are subdued by the hegemony of science and technocratic management. Devoid of actual local voices, the 'experts and elites' in the Northeast region shape biased conservation discourses and strategies that do not reflect reality. Unless there is support or approval from local citizens, wildlife conservation will continue to be a story of science, state, and wildlife activism.

## Conclusion

The absence of social sciences and humanities in conservation biology or wildlife management is not new. Scientists trained in wildlife sciences are trained to implement conservation projects where aspects of biological sciences dominate. For example, conservation activities are mainly focused on quantifying forest cover, counting animal populations, and measuring threats to the health of the ecosystem. These projects are shaped with little or no insights from the social history of the landscapes they operate in.

Sustainable hunting policies can be developed if they have been co-produced through engagement with local communities. Local institutions, within both the government and non-governmental sectors, need to be involved, ideally comprising members from local communities at both planning and implementation levels. Such decentralised conservation strategies could use inputs from academia to help build local-suited sustainable use strategies.

In north-east India, where wildlife hunting is still considered a part of the social-cultural ethos, any form of conservation and its effectiveness needs more scrutiny. The challenge is to find a middle ground where conservation can run hand in hand with communities' aspirations. There is a need for a holistic understanding of hunting and conservation. Wildlife biologists, social scientists, conservation practitioners, and those from the development sector can and should work together to find solutions that will conserve species important for subsistence economies, diet, and culture. Studying hunting only from an ecological point of view is a severe limitation, and there is scope to include other aspects such as market linkages and changes in socio-economic aspects that may provide a start to understanding the complexities that wildlife hunting involves in north-east India.

## References

Aisher, A. (2007). Voices of uncertainty: Spirits, humans and forests in upland Arunachal Pradesh, India. *South Asia: Journal of South Asian Studies*, *30*(3), 479–498. DOI:10.1080/00856400701714088

Aisher, A., & Damodaran, V. (2016). Introduction: Human-nature interactions through a multispecies lens. *Conservation and Society*, *14*(4), 293–304. www.jstor.org/stable/26393253

Aiyadurai, A. (2011). Wildlife hunting and conservation in Northeast India: A need for an interdisciplinary understanding. *International Journal of Galliformes Conservation*, 2, 61–73.

Aiyadurai, A. (2016). 'Tigers are our brothers:' Understanding human-nature relations in the Mishmi Hills, Northeast India. *Conservation and Society*, *14*(4), 305–316. DOI:10.4103/0972–4923.197614

Aiyadurai, A. (2018). Human-animal relations: A view from the Mishmi hills. *Seminar*, *702*(2), 1–8.

Aiyadurai. A., & Banerjee, S. (2019). Bird conservation from obscurity to popularity: A case study of two bird species from Northeast India. *GeoJournal*, *85*(4), 901–912. https://doi.org/10.1007/s10708-019-09999-9

Aiyadurai, A., Singh, N. J., & Milner-Gulland, E. J. (2010). Wildlife hunting by indigenous tribes: A case study from Arunachal Pradesh, north-east India. *Oryx*, *44*(4), 564–572. https://doi.org/10.1017/S0030605309990937

Aiyadurai, A., & Velho, N. (2018). The last hunters of Arunachal Pradesh: The past and present of wildlife hunting in North-east India. In N. Velho & U. Srinivasan (Eds.), *Conservation from the margins* (pp. 69–93). Hyderabad: Orient Blackswan.

Banerjee, S. (2016). *Report on traditional hunting practice, wild meat consumption and wildlife trade in Nagaland, India*. Cambridge: TRAFFIC.

Beck, U. (2010). Climate for change, or how to create a green modernity? *Theory, Culture & Society*, 27(2–3), 254–266. https://doi.org/10.1177/0263276409358729

Bhupathy, S., Kumar, S. R., Thirumalainathan, P., Paramanandham, J., & Lemba, C. (2013). Wildlife exploitation: A market survey in Nagaland, North-Eastern India. *Tropical Conservation Science*, 6(2), 241–253. https://doi.org/10.1177/194008291300600206

Blench, R. (2016). *The evolution of hunting among the Idu, a people of Arunachal Pradesh*. www.rogerblench.info/Language/NEI/Mishmi/Idu/Iduanth/Idu%20hunting.pdf

Chutia, P. (2010). Studies on hunting and the conservation of wildlife species in Arunachal Pradesh. *SIBCOLTEJO*, *5*, 56–67.

Das, A. K., & Shukla, S. P. (2007). Meeting Report: Biodiversity and indigenous knowledge system. *Current Science*, *92*(3), 275–276. www.jstor.org/stable/24096713

Datta, A. (2007). Protecting with people in Namdapha: Threatened forests, forgotten people. In G. Shahabuddin & M. Rangarajan (Eds.), *Making conservation work: Securing biodiversity in this new century* (pp. 165–209). New Delhi: Permanent Black.

Datta, A., Anand, M. O., & Naniwadekar, R. (2008). Empty forests: Large carnivore and prey abundance in Namdapha National Park, north-east India. *Biological Conservation*, *141*(5), 1429–1435. https://doi.org/10.1016/j.biocon.2008.02.022

Datta Roy, A. (2018). *Swidden, hunting and Adi culture: Highland transitions in Arunachal Pradesh, India*. [Doctoral thesis, Manipal Academy of Higher Education]. Shodhganga.

D'cruze, N., Singh, B., Mookerjee, A., Harrington, L. A., & Macdonald, D. W. (2018). A socio-economic survey of pangolin hunting in Assam, Northeast India. *Nature Conservation*, *30*, 83–105. https://doi.org/10.3897/natureconservation.30.27379

Hilaluddin, K. R., & Ghose, D. (2005). Conservation implications of wild animal biomass extractions in northeast India. *Animal Biodiversity and Conservation*, *28*(2), 169–179.

Janaki, M., Pandit, R., & Sharma, R. K. (2020). The role of traditional belief system in conserving biological diversity in the Eastern Himalaya Eco-region of India. *Human Dimensions of Wildlife*, *26*(1), 13–30. https://doi.org/10.1080/10871209.2020.1781982

Knight, J. (Ed.). (2004). *Wildlife in Asia: Cultural perspectives*. New York: Routledge. https://doi.org/10.4324/9780203641811

Kothari, A., Pathak, N., & Vania, F. (2000). *Where communities care: Community-based wildlife and ecosystem management in South Asia*. Pune: Kalpavriksh and International Institute for Environment and Development.

Krishnan, P. R., Ramakrishnan, R., Saigal, S., Nagar, S., Faizi, S., Pawar, H. S., Singh, S., & Ved, N. (2012). *Conservation across landscapes: India's approaches to biodiversity governance*. United Nations Development Programme. www.in.undp.org/content/dam/india/docs/EnE/conservation-across-landscapes.pdf

Kumar, A., & Riba, B. (2015). Assessment of effectiveness of conservation action adopted for hornbill species in Arunachal Pradesh, India: The Great Indian Hornbill (*Buceros bicornis*). *International Journal of Conservation*, 6(1), 125–134.

Morris, B. (1998). *The power of animals*. Oxford: Berg. https://doi.org/10.4324/9781003087151

Myers, N., Mittermeier., R. A., Mittermeier, C. G., da Fonseca, G. A., & Kent, J. (2000). Biodiversity hotspots for conservation priorities. *Nature*, *403*(6772), 853–858. DOI: 10.1038/35002501

Naro, E., Mero, E. L., Naro, E., Kapfo, K. U., Wezah, K., Thopi, K., & Chirhah, T. (2015). Project hunt: An assessment of wildlife hunting practices by local community in Chizami, Nagaland, India. *Journal of Threatened Taxa*, 7(11), 7729–7743. DOI:10.11609/jott.2317.7729–7743

Nature Conservation Foundation-India. (n.d.). *Hornbill nest adoption program*. www.ncf-india.org/eastern-himalaya/hornbill-nest-adoption-program

Nijhawan, S. (2018) *Human-animal relations and the role of cultural norms in tiger conservation in the Idu Mishmi of Arunachal Pradesh, India*. [Doctoral thesis, University College London]. UCL Discovery.

Nijhawan, S., & Mihu, A. (2020). Relations of blood: Hunting taboos and wildlife conservation in the Idu Mishmi of Northeast India. *Journal of Ethnobiology*, *40*(2), 149–166. https://doi.org/10.2993/0278-0771-40.2.149

Pathak, N. (Ed.). (2009). *Community conserved areas in India – A directory*. Kalpavriksh: Pune/New Delhi. https://kalpavriksh.org/wp-content/uploads/2019/01/Community-Conserved-Areas-in-India.pdf

Rahman, A. P. (2016, December 7). Hunters turn protectors of threatened hornbills in northeast India. *The Third Pole*. www.thethirdpole.net/2016/12/07/hunters-turn-protectors-of-threatened-hornbills-in-northeast-india/

Rane, A., & Datta, A. (2015). Protecting a hornbill haven: A community-based conservation initiative in Arunachal Pradesh, Northeast India. *Malayan Nature Journal*, *67*(2), 203–218.

Selvan, M, K., Veeraswami, G. G., Habib, B., & Lyngdoh, S. (2013). Losing threatened and rare wildlife to hunting in Ziro valley, Arunachal Pradesh, India. *Current Science*, *104*(11), 1492–1495.

Sethy, J., & Chauhan, N. P. S. (2012). Conservation status of Sun Bear (*Helarctos malayanus*) in Nagaland state, North-East India. *Asian Journal of Conservation Biology*, *1*(2), 103–109.

Singh, R. K., Alves, R. N., & Ralen, O. (2014). Hunting of Kebung (*Ratufa bicolor*) and other squirrel species from Morang Forest by the Adi tribe of Arunachal Pradesh, India: Biocultural conservation and livelihood dimensions. *Regional Environmental Change*, *14*(4), 1479–1490. DOI:10.1007/s10113-014-0590-3

The Energy and Resources Institute. (2015). *Documentation of community conservation areas in Nagaland*.

Tynsong, H., Tiwari, B. K., & Dkhar, M. (2012). Bird hunting techniques practised by War Khasi community of Meghalaya, North-East, India. *Indian Journal of Traditional Knowledge*, *11*(2), 334–341. http://nopr.niscair.res.in/handle/123456789/13866

Velho, N., Karanth, K. K., & Laurance, W. F. (2012). Hunting: A serious and understudied threat in India, a globally significant conservation region. *Biological Conservation*, *148*(1), 210–215. doi:10.1016/j.biocon.2012.01.022

Velho, N., & Laurance, W. F. (2012). Hunting practices of an Indo-Tibetan Buddhist tribe in Arunachal Pradesh, North-East India. *Oryx*, *47*(3), 389–392. DOI: 10.1017/S0030605313000252

# 9 Sustainable grazing practices

## Conserving biodiversity in an Asian tropical grassland

*Pankaj Joshi*

### Introduction

The Banni, one of Asia's largest tropical grasslands, is home to the Maldharis, a pastoral community. Spread across 2,500 square kilometres, this region is home to a variety of flora and fauna (Figure 9.1). The Maldharis rear buffaloes and cows in this landscape, which they share with the caracal, wolf, chinkara, spiny-tailed lizard, desert cat, desert fox, Houbara bustard, and many migratory waterfowl and birds. The long-term sustainability of landscape-based livelihoods depends on maintaining the ecological integrity of the landscape. For instance, an adequate population of ground nesting birds, reptiles, mammals, insects, etc. is necessary for maintaining native grassland habitats. Many pastoralists in the Banni region are aware of the local animals, birds, insects, their habitats, and their significance to the functioning of the environment, although this knowledge is rarely documented.

In order to understand the management of grazing areas, the Research and Monitoring in the Banni Landscape (RAMBLE) – an open research platform established through multi-institutional collaboration – undertook a participatory resource mapping exercise to map the seasonal and traditional grazing corridors to inventory the biodiversity of the region. The project focused on 13 of the 19 villages in Banni. Village panchayats and their Community Forest Management Committees (CFMCs) were involved in developing a participatory action plan for restoration of grazing resources and related biodiversity. This chapter presents a case study of one village panchayat, Daddhar (Figure 9.2), as it developed participatory protocols for conservation of biodiversity and restoration of native ecosystems.

### Participatory conservation planning

The restoration of native ecosystems requires the protection and conservation of microhabitats on which the existing biodiversity of the area rely. In order to prepare a biodiversity conservation plan, the project needed

DOI: 10.4324/9781003343493-12

*Figure 9.1* Tropical grasslands of Banni in Kutch, Gujarat

to create a dataset or biodiversity register. The project estimated that such data would enable the creation of linkages between nature resources management and the livelihoods and sustenance of village communities. Local communities are imminently capable of mapping species and ecosystem biodiversity, and monitoring, conserving, and managing natural resources, which are directly related to their livelihood. The project was developed as a pilot study to understand land-use planning for biodiversity conservation and livelihood security at the village panchayat level in Banni, and other communal grazing areas of the District.

Village residents in Kutch actively participated and jointly developed regulations for biodiversity conservation. For example, Guneri in Lakhpat Taluka of Kachchh District decided to enforce the traditional system of monitoring mangroves. Villagers were knowledgeable about cutting the main branches of the local mangrove varieties, allowing for the trees to regenerate. Sayra in Nakhtrana Taluka resolved to protect the White-naped Tit in its native habitats by reducing tree cutting and other pressures. They also developed sign boards specifying restricted activities. Lathedi in Lakhpat Taluka developed regulations restricting grazing in areas that are the habitat of the herb, *Olax nana*.

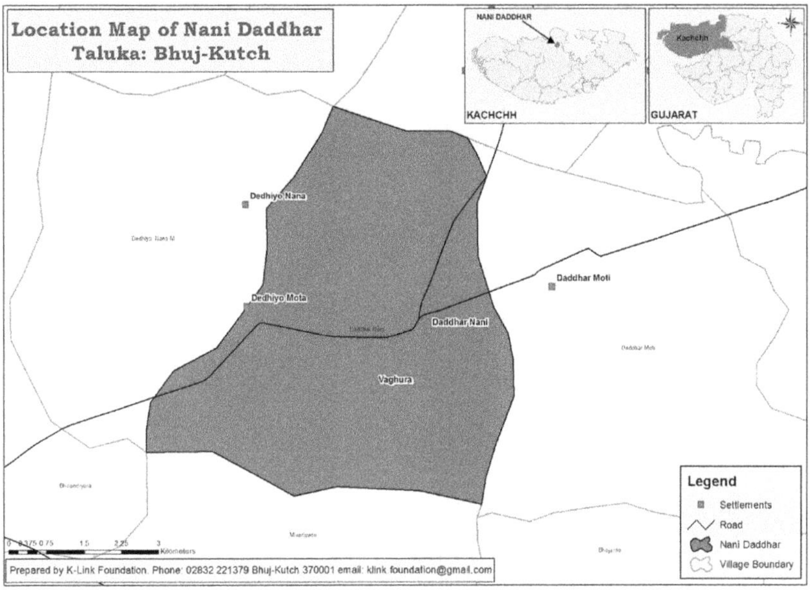


*Figure 9.2* Location of Nani Daddhar Taluka
*Map: K-Link Office, Bhuj*

## Area profile and study

### *Banni grasslands*

Banni is spread over 48 hamlets that are organized into 19 panchayats that comprise 7,500 families. Communities of this region are famous for their craft work-embroidery, using materials such as leather and wood (Virmani et al. 2010). The Banni grasslands are interspersed with wetlands making it a unique ecosystem, which is bound on the southern side by the Great Rann. In the winters – from November to March, marshy grasslands with water-logged sedges are home to thousands of migratory species. In the dry period beginning from April, the area quickly turns into savannas, with plants that are able to tolerate the shift from wetland to dryland.

The clay-rich coarse soils of the Banni support open grassland and savanna woodlands. The understory is comprised of saline tolerant and intolerant herbs, grasses, and shrubs. An estimated 17 tree species are also part of the ecosystem. *Ochtochloa compressa* (Madhanu, Gandhiro), *Suaeda nudiflora* (Lano), *Suaeda fruticosa* (Unt morad), *Cressa cretica* (Oin), *Sporobolus marginatus* (Dhrabad), *Chloris barbata* (Siyar Puchha), and *Aeluropus lagopoides* (Lolar) are the common saline tolerant plant species that are dominant throughout in Banni. (Patel and Joshi 2011; Thorat et al. 2019).

*Figure 9.3* Focus group discussion in Daddhar
*Photo: Ramesh Bhatti*

## *Methods and approach*

Seven focus group discussions (FGDs) were held with the pastoralist communities of Daddhar Panchayat comprising the villages of Nani Daddhar (Figure 9.2), Moti Daddhar, Vaghura, Sadhara, and Dedhiya (Figure 9.3), in addition to which field-based surveys were undertaken throughout the landscape to produce a base map of grazing patches. These maps drew on the knowledge of the pastoralist communities on various aspects of grassland ecology, biodiversity, and sustainable traditional grazing practices.

Google Earth was used to map grazing patches, alongside the use-pattern analysis of the landscape of Daddhar Panchayat that was representative of almost all habitat categories commonly found in the Banni region. This landscape comprised numerous small to medium sized (between 1 and 6 km) and a few large (more than 10 km) inter-mixed patches of a variety of vegetational compositions, which made the mapping task quite challenging.

In addition, a reconnaissance of the Daddhar Panchayat was carried out to identify landmarks which could operate as control points for carrying out ground-truthing to develop vegetation and composition maps through

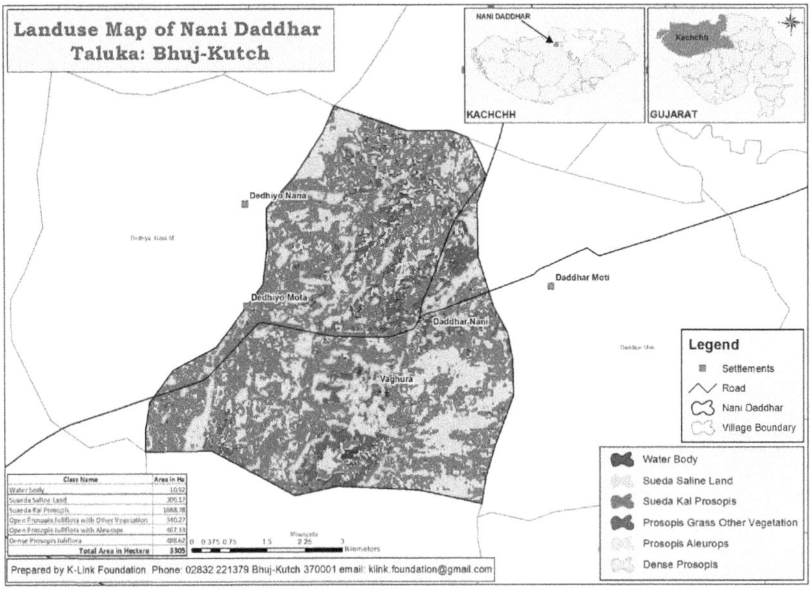


*Figure 9.4* Land use and vegetation types of Nani Daddhar Taluka
*Map: K-Link Office, Bhuj*

available satellite data with K-Link. The major habitats of this landscape are the *Prosopis* trees mixed, sedge and grasses, bush vegetation (*Suaeda* spp., *Prosopis* mixed), saline patches (*Suaeda* sp. mixed), and water bodies (Figure 9.4).

The distribution of plant species has been recorded for these habitats along with other data, such as dominant vegetation, direct and indirect evidence of biodiversity, density of *Prosopis*, water availability, and soil types (Figure 9.5). Subsequent visits were also made for intensive field work that involved collecting data on vegetation composition in a total of 67 circular plots distributed equally in various dominant habitats: *Prosopis* Dense, *Prosopis* Sparse/medium, *Prosopis* Open with *Suaeda*, Saline land/with *Suaeda, Virda* Dense Vegetation/Pvt. Grass plot, *Acacia nilotica* patches/wetland, and *Cyperus-Scirpus* & *Cress* habitat (Figure 9.6). Additionally, unsupervised and supervised image classifications will be done to calculate the spread of these habitats in each of the traditional grazing patches of Daddhar Panchayat.

### *Distribution of grazing patches and dominant grasses*

Based on FGDs, 50 to 55 traditional grazing patches were identified in Daddhar Panchayat, which were historically used for the communities' livestock

*Figure 9.5* Prosopis mixed saline habitat in Daddhar

*Figure 9.6* Ground-truthing expedition

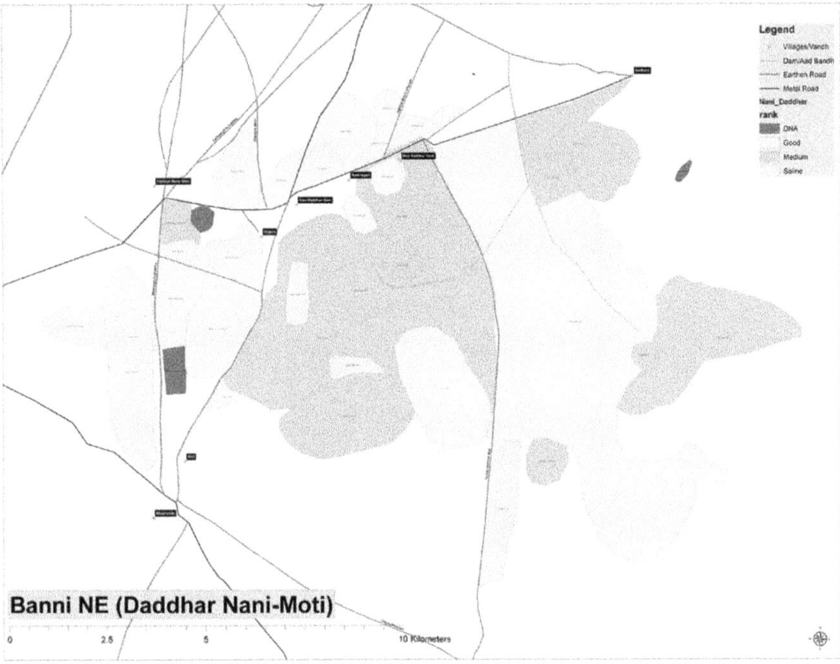


*Figure 9.7* Distribution of grazing areas
*Map: K-Link Office, Bhuj*

grazing. Figure 9.7 shows the distribution of various grazing patches with the seasonal grazing pattern of livestock of dependent villages (i.e. nine villages located in the north, 14 from the east, six from the south, and seven from the west). Most of the respondents expressed that their livestock used to migrate to grazing patches far away from the villages during summer seasons (e.g. *Kunjiwaro Chacha, Hanj Talav, Vaghurawaro Chachha, Kagavari Sim*), while during the monsoon, they grazed in patches (e.g. *Tango, Mitho Bhon, Saiyad Malu Kabrastan Area, Laiwaro Bhon*) closer to the villages due to greater availability of palatable grasses. Only two (*Fatel Talav* Area and *Juno Talav* Area-Vaghura) and three (*Keradwari Mori* Area, Plot Area, *and Sayabwaro Zil* Area) patches are grazed during winter, and winter and summer, respectively (Figure 9.7).

The traditional name of each grazing area indicates the dominant vegetation characteristics, soil and salinity type, water availability, distribution of existing biodiversity, and vegetation density. For instance, *Kerad Wari Mori* is named for Kerad, a local name of *Capparis decidua* and Mori, meaning slightly saline area. *Lai waro Bhon* – Lai is local name of *Tamarix* sps. and Bhon means slightly depressed area. *Aakk waro Thath* – Aakk is local name of *Calotropis procera* and Thath means medium water table area. *Kandhi*

*Figure 9.8* Hanj Tal

*waro Kado* – Kandhi is local name of *Prosopis cineraria* and Kado means dense vegetation with narrow line. *Kunjiwaro Chacha* – Kunj is the local name of Cranes and Chhachh is a shallow water table area. *Hanj Tal* – Hanj is local name of Flamingos and Tal means shallow water body (Figure 9.8). *Bhagad Chhachh* – Bhagad is local name of Wolf and Chhachh is shallow water table area. *Shiyad waro todo* – Shiyad is local name of Jackle and todo means dense vegetation area.

In addition, based on FGDs, the study was able to categorize and rank grazing patches into three categories (i.e. good, medium, and saline) based on their soil profile, existing vegetation composition, distance from water bodies and villages. Interestingly, three grazing patches on the southern side (*Kabulvaro* Plot Area, *Ecology Plot* Area *and Kakkavera Sim*) are classified as high-to-medium saline by pastoralist communities. *Suaeda-Cress-Aleropus* (Lano-Oin-Khariyu) are one of the indicator plant species of this category. More than 17 grazing areas (almost 50% of the total landscape of village panchayat) are classified under the good category and they are very close (less than two kilometres) to villages and water bodies. These areas are dominated by a mix of grasses, including high nutritive grasses, high fibre-containing cellulose grasses, and saline mixed grasses required for livestock.

All the areas categorized as 'good' record high numbers of species including mammals, reptiles, and birds (including migratory species). However, more than 50% of these 'good' areas are undergoing a *Prosopis* invasion. Hence, these grazing patches are high priority for village panchayats, in terms of biodiversity conservation and restoration of native grassland, in order to sustain their milk economy. Following on from the study, six grazing patches (*Tango Sim, Laiwaro Bhon, Khatuwari Khan, Kharo Bhon, Vanwari Sim*, and *Denovalia Sim*) have been identified for conservation and restoration, as these grazing patches are available in all seasons, and made use of by most villages in the study area.

### *Distribution of biodiversity*

Banni is endowed with a high diversity of wetland birds (54 species), terrestrial birds (95 species), 13 species of reptiles, and 12 species of mammals (GUIDE, 2011), in addition to more than 190 plant species (Patel and Joshi 2011). As understood from FGDs, the main mammal species are Blue Bull (*Roz, Nil Gay*), Chinkara (*Haran*), Hare (*Sasla*), Jackal (*Shiyar*), Wolf (*Bhagad*), Hyena (*Charak, Zarakh*), Desert and Indian Foxes (*Lokadi*), Jungle Cat (*Mini*), and Caracal (*Hanotro*). Herpetofauna exist in various habitats of Daddhar Panchayat and include spiny-tailed lizard (*Sandho*), common monitor lizard (*Go*), and garden lizard (*Kachindo*). During a good rainfall year, the water bodies such as *Kunjiwaro Chacha*, *Hanj Tal*, and *Vaghurawaro Chachha* are important staging grounds for thousands of migratory cranes and over 20 to 25 species of migratory and other resident birds.

Based on FGDs, field surveys, and available literature on Daddhar Panchayat and adjoining areas, base maps were produced on the distribution of key species in each grazing patch (Figure 9.9). These maps identified potential ecologically significant areas (ESA), including *Kunjiwaro Chacha, Hanjvaro Talav, Keradvari Mori, Vanvari Vendh, Tango Sim,* and *Laiwaro Bhon.* These patches had high recorded numbers of diverse species and abundance, including native trees, shrubs, grasses, mammals, reptiles, and birds. In addition, due to the availability of potential food and the existence of wetlands adjoining these areas, there is scope for the ESA-recommended patches to be prioritized for conservation and management following a participatory approach.

## Strategies for establishing and maintaining native biodiversity

Globally, domestic livestock grazing is seen as the root cause for change in species composition and associated loss of biodiversity richness in grasslands. Shifts in species composition frequently involve the replacement of palatable species by unpalatable species. Once unpalatable species attain dominance on the landscape, it can be difficult to reverse the change by

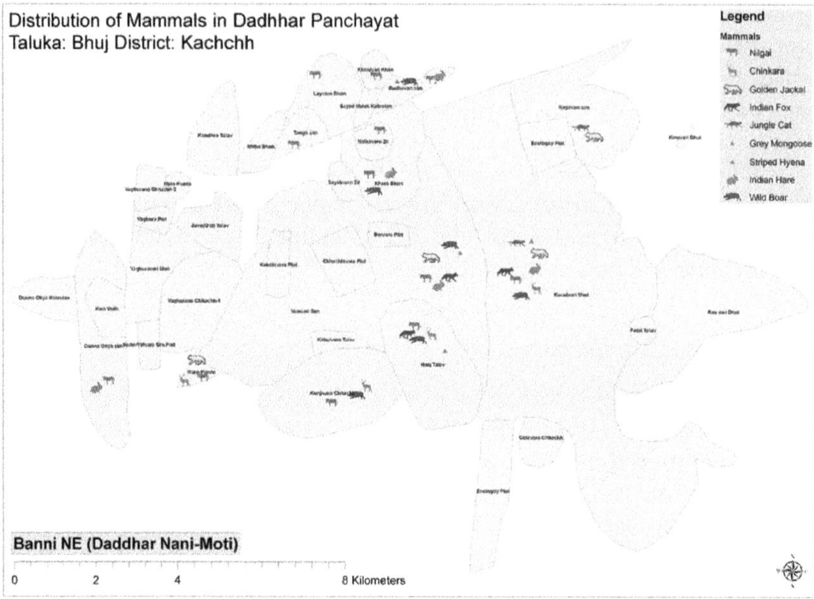


a

*Figure 9.9* Distribution of wildlife in the grazing areas

*Map: K-Link Office, Bhuj*

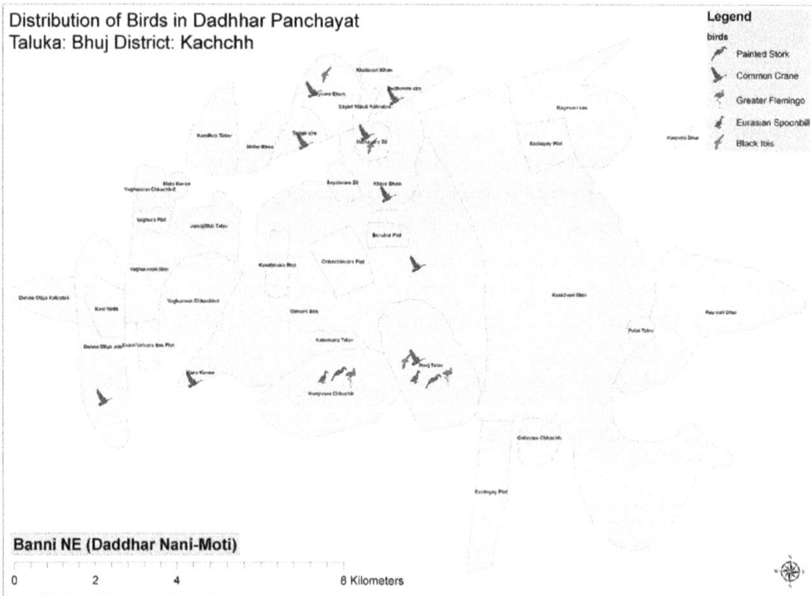


b

*Figure 9.9* (Continued)

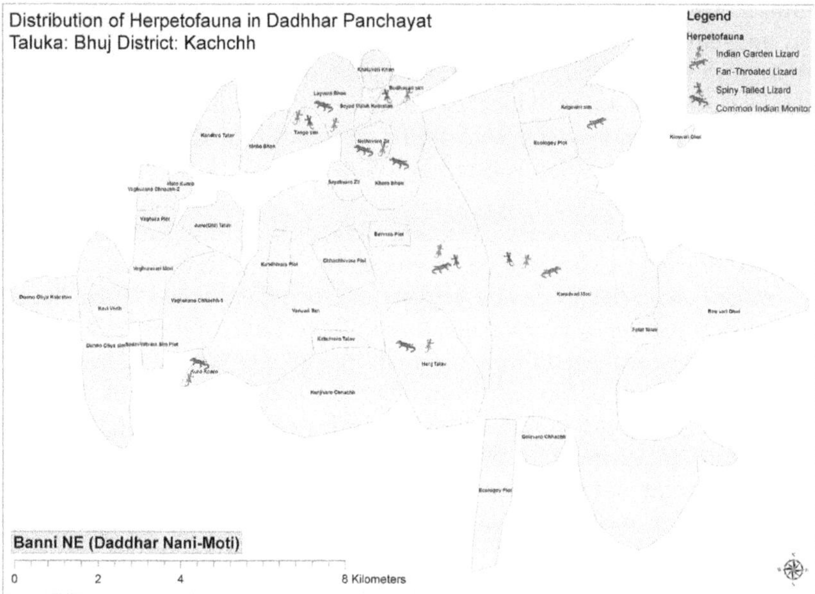


c

*Figure 9.9* (Continued)

relaxing or even removing grazing (Distel et al. 2008). In the past, due to the high intensity of grazing by immigrant livestock in Banni, native biodiversity degraded, severely affecting grasslands (Joshi et al. 2009). In light of this, recommendations to restore and maintain the native grassland habitats with significant floral and faunal components are vital.

Discussions with pastoralist communities revealed that they are not in favour of reducing the herd size of less productive livestock, despite comprehension that in the long term it would pose a threat to the landscape. The effects of selective grazing and trampling needed to be considered when exploring alternative strategies for management and conservation. RAMBLE is currently conducting further research in order to produce strategic directions to establish and maintain the native landscape. Directions are particularly important to develop methods for grazing management practices that increase the abundance of desirable species that are still prevalent, and for sowing native grasses in areas where the most desirable species are no longer present.

## Participatory management plan for *Prosopis juliflora*

It is well established that domestic animals such as cows, buffaloes, goats, and horses feed on the pods of *Prosopis* and disperse them to other adjoining habitats through excretory material during the process of grazing, thus contributing to the expanding range of this species across many habitats. In addition, Karthik et al. (2009) recorded that wild animals such as jackal and wild boar are also responsible for the dispersal of the seeds of *Prosopis* in Banni. With this in mind, a management plan for *Prosopis* was outlined that relies on the involvement of local communities that are involved in Prosopis-based charcoal-making. The level of involvement of the local community in different phases of the plan is listed as follows:

- *Prosopis juliflora* biomass extraction methods from the selected and shortlisted grazing patches will be developed by CFMC through its authority granted under the *Scheduled Tribes and Other Traditional Forest Dwellers (Recognition of Forest Rights) Act, 2006*. Collection of Prosopis biomass will be managed through a well-organized structural framework to avoid illegal use of biomass from other native trees and shrubs. In parallel, local pastoralist communities will be involved in restoring the potential grazing patches with other native wild tree species, through habitat improvement and biodiversity restoration programmes.
- It is further proposed that an institutional framework with control mechanisms be developed by CFMC under available central and state government schemes for community activities. This could involve captive plantations on adjoining panchayat lands identified by CFMC. Support might also take the form of tools, vehicles for transportation, and assistance through national rural employment guarantee schemes, microfinance credit facilities, etc. that may be made available to local charcoal-making communities.

## Conclusion

Grasslands are important, biodiverse ecosystems in drylands. The consequences of losing the composition of native grasses can have serious implications for the ecosystem health and functioning of grasslands, which could lead to the loss of species diversity. It is obvious that loss of grassland species has economic implications. The livelihoods and cultural heritage of pastoralists will be endangered if grasslands and grassland resources are further depleted. Findings of the pilot study generated insights that are relevant for the management and restoration of grasslands in Banni and in similar landscapes.

A key finding is that the migration of pastoralist communities with their livestock is an important strategy to access seasonal resources in all the panchayats of Banni, including Daddhar. This strategy has also helped to maintain the productivity and regeneration capacities of the land. Therefore, to maintain and manage the uniqueness of native habitats with a composition of fodder species, the establishment of an open pasture land with minimum disturbance of soil strata and other existing biodiversity is critical.

Furthermore, based on the literature, the study did uncover degradation to some grasslands due to over-grazing. This requires the development of sustainable utilization strategies that may be applied to migrating grazing communities from adjoining panchayats, districts, and states who seek to access grassland resources in a particular landscape. A participatory mechanism or community-based system to control and manage non-local livestock may prevent unrestrained grazing in an area. The overall goal is to achieve a balance among increased potential grazing areas, sustainable use of natural resources, and the restoration–conservation of biodiversity through reversing land degradation. Such efforts have enormous potential to enhance local livelihoods and reduce the precarity of pastoral lives.

## References

Distel, R. A., Pietragalla, J., Iglesias, R. M. R., Didoné, N. G., & Andrioli, R. J. (2008). Restoration of palatable grasses: A study case in degraded rangelands of central Argentina. *Journal of Arid Environments*, *72*(10), 1968–1972.

Joshi, P. N., Kumar, V., Koladiya, M., Patel, Y. S., & Karthik, T. (2009). Local perceptions of grassland change and priorities for conservation of natural resources of Banni, Gujarat, India. *Frontiers of Biology in China*, *4*(4), 549–556. DOI:10.1007/s11515-009-0041-6

Karthik, T., Patel, Y., Koradiya, M., Pardeshi, M., & Joshi, P. N. (2009). Dispersal of Prosopis juliflora seeds in the feces of wildlife in the Banni Grassland of Kachchh Desert, Gujarat. *Tigerpaper*, *36*(3), 31–32.

Patel, Y. S., & Joshi, P. N. (2011). A floristic inventory in the Banni Region of Bhuj Taluka, Kachchh District, Gujarat (India). *The Indian Forester*, *137*(9), 1114–1121.

Thorat, O. H., Nerlekar, A. N., & Joshi, P. N. (2019). *Grasses of Banni (bilingual)*. Pune: BAIF Development Research Foundation.

Virmani, S., Das, S., & Bhatti, R. (2010, February 1–4). *Pastoral life style in Banni Region: A case of conservation of Banni buffalo, grassland ecosystem and utilization of commons* [Paper presentation]. New Delhi: International Buffalo Conference.

# Part IV

# Intangible benefits

# 10 Counting to conserve

## The role of communities and civil society in monitoring marine turtles

*Kartik Shanker and Muralidharan Manoharakrishnan*

### Introduction

Evolutionary processes take place over long-time scales (thousands or millions of years) while ecological processes take place over shorter time periods (tens or hundreds of years) (Carroll et al. 2007). Even so, many ecological changes occur over periods of time longer than the lifetime of projects or even people. On the one hand, the processes themselves may be slow – long-lived, late maturing plants and animals take decades to grow and fully mature. On the other hand, ecological systems are rife with inherent variation, and this may result in fluctuations from year to year that are more a product of chance than of any trend. In such systems, most parameters are hard to measure precisely, adding to the errors or variation in their estimation. This means it may take years or decades of observation to detect a pattern or trend in populations and communities. Detecting these trends is critical to making decisions about conservation, management, or sustainable resource use.

Most studies, however, are carried out within short periods of time due to a combination of financial, logistic, and personnel constraints. Most projects are funded for short durations and often require quick results to obtain continued support, in terms of both money and logistics. Most research personnel are engaged in projects that necessitate a short-term commitment such as a masters, PhD, or other research programme. Hence, a large majority of ecological projects do not incorporate a temporal axis of change.

The idea of long-term ecological monitoring has gained importance in recent years, especially in the context of climate change, with several national and international networks that seek to collect data on various physical (climate, soil) and biological (fauna and flora) parameters for the foreseeable future. These include programmes such as the long term ecological research (LTER) networks in USA and Europe, the National Ecological Observatory Network (NEON) in USA, the global Nutrient Network (NutNET), and so on (Mirtl et al. 2018). The benefit of institutionalizing such monitoring is that it can then go beyond the timeframes of individual projects and researchers.

DOI: 10.4324/9781003343493-14

Many initiatives such as citizen science also engage with civil society to generate long-term datasets. Of course, expanding the personnel involved in such data collection allows for greater spatial and temporal coverage. But these programmes go far beyond merely increasing the number of hands and eyes available for data collection. The objectives of citizen involvement extend to engaging communities, educating society, and building a connection with nature and the environment (Dickinson et al. 2010). Hence, the motivation for the involvement of participants is more than just data collection, as these programmes can garner support of a wider swathe of the community over longer periods of time.

As long-lived, late maturing, migratory species, marine turtles require long-term monitoring to gain insights into their population biology and trends. There are some remarkable long-term monitoring programmes for marine turtles that rank among the first such programmes in the world. Archie Carr, considered the 'father of sea turtle biology', initiated his research on green turtles in Tortuguero in the 1950s (Carr 1967), and the monitoring programme he subsequently started carries on into the present day (Troëng and Rankin 2005).

Similarly, a monitoring programme was initiated for loggerhead turtles at Mon Repos, Queensland by Colin Limpus and Robert Bustard in the 1960s (Shimada et al. 2016). In South Africa, George Hughes introduced a monitoring programme for loggerhead and leatherback turtles around the same time (Robinson et al. 2018), and George Balazs and his team began their programme in the Hawaiian islands to monitor the then threatened '*Honu*' or green turtles, which have seen a remarkable resurgence in their population since (Balazs and Chaloupka 2004). These programmes have provided extensive data to study trends, catalysed conservation efforts, and engaged civil society and communities.

Sea turtle conservation in India began in the early 1970s, with differing approaches, in Odisha (Orissa) and Chennai (Madras). The rich history of these efforts, and others that were initiated, provide insights into the challenges of and opportunities in such programmes. Here, we explore the ways in which communities and civil society engage in such long-term programmes, with outcomes for both research and conservation.

## Keeping track of turtles

Sea turtles are amazing animals – migrating thousands of kilometres across the open ocean between feeding and breeding grounds and using the earth's magnetic field to navigate and return to the beaches where they were born (Carr 1967; Lutz et al. 2002)). Some, such as the leatherback turtle, dive to depths of over 1,000 metres, and foray into cold temperate waters in search of jellyfish (Spotila and Tomillo 2015). Others, such as the green turtle, feed exclusively on seagrass and algae, while hawksbill turtles feed on sponges (Lutz et al. 2002).

Sea turtles spend their entire life at sea, coming ashore only to nest as adults. Females come ashore and lay 100–150 eggs in a two to three foot-deep nest above the high tide line. The eggs incubate for about two months and hatchlings emerge synchronously at night and return to the sea immediately, locating it by orienting to the brighter horizon, the reflection of the moon and stars on the water. They live in the open oceans for several years floating around in seaweed rafts on transoceanic currents and gyres, until they transition to their sub-adult and adult feeding habitats (Bjorndal 1995).

Among the most remarkable of the natural phenomena exhibited by sea turtles is the synchronized mass nesting of the two ridley species (or *arribada*, meaning arrival in Spanish) – the olive ridley which is found throughout the world, and the Kemp's ridley which nests only on the east coast of Mexico (Plotkin 2007). During this event, thousands or tens of thousands of ridley turtles come ashore at night, and nest on small stretches of beach in central America and in Odisha on the east coast of India. Night after night, the beach is filled with turtles, digging nests and laying eggs, until they are accidentally digging up the nests of their predecessors and destroying them in the process. Mass nesting can go on for a few days, and by the end, several hundred thousand turtles may have nested (Plotkin 2007).

In recent decades, there has been much consternation about the decline of sea turtle populations worldwide. Therefore, a call has been sent out to monitor these populations so that one can determine whether they are increasing, decreasing, or stable. This, however, is not straightforward for several reasons (Hamann et al. 2010, Rees et al. 2016).

The life history of sea turtles provides both challenges and opportunities (Bjorndal 1995). The fact that they come ashore to nest provides an opportunity to count them on an annual basis. However, it may take years for impacts such as adult mortality to reflect in nesting trends. Moreover, there is tremendous inherent variation in nesting. Sea turtles do not breed every year, many will breed once in two or three years. This means that, out of all the adults in the feeding grounds, a different proportion may migrate to the breeding ground each year. This itself will cause variation in the nesting population size each year. Given these factors, it becomes necessary to monitor these populations for many years or even decades. Their long-distance migrations create challenges in terms of determining their habitat use and where they might need protection. Now, satellite transmitters are used to track these animal movements from breeding to feeding ground and vice versa (Jeffers and Godley 2016). Modern tools developed in molecular genetics have also allowed genetic mapping of breeding to feeding grounds.

Here, we provide two case studies from India, one involving the monitoring of sea turtles on land – the mass nesting population of olive ridleys in Odisha, and the other involving monitoring sea turtles at sea – the tracking of leatherback turtles from the Andaman Islands using satellite telemetry. Both projects have depended heavily on a cadre of trained and highly motivated field staff from the local communities. We also describe an initiative for monitoring sea turtle populations at a broader national scale through a partnership with local organizations, as well as attempts at turtle tourism.

## Counting turtles in Odisha

As spectacular as the sight may be, a mass nesting event of ridley turtles presents a conundrum for researchers. How does one count tens of thousands of turtles at a time? Clearly, it is not possible to count every single one (Gates et al. 1996). However, given the prevalence of a variety of anthropogenic threats and possible declines, it becomes important to do so.

The mass nesting population of olive ridley turtles in Odisha was brought to the attention of the scientific community in 1974 by H. Robert Bustard, an FAO consultant working on a salt-water crocodile project. In an article in the journal Tigerpaper, Bustard suggested that Gahirmatha was the world's largest olive ridley rookery (Bustard 1976). He set up a research programme for crocodiles and sea turtles, and his fellow turtle researcher, the late Chandra Sekhar Kar, went on to study these animals for the next decade. During this time, Kar discovered a second mass nesting beach at the Devi River mouth (Kar 1982), although the site has not had mass nesting since the late 1990s (Shanker et al. 2004). In 1994, another young researcher from the Wildlife Institute of India, Bivash Pandav discovered a third mass nesting beach at Rushikulya River mouth, in Ganjam District on the southern Odisha coast (Pandav et al. 1994).

How this beach was brought to Pandav's attention is of significance. Pandav and Kar had met an enterprising fisherman during their surveys of the Odisha coast and he sent them a postcard informing them of a mass nesting event several months after their visit. The fisherman, the late Damburu Behera, would go on to become the mainstay of Pandav's programme, and the research and monitoring that would follow. During the course of his PhD in the 1990s, Pandav set up monitoring camps at Gahirmatha, Devi River Mouth, and Rushikulya. All these camps employed local staff to monitor the beaches, maintain the camps and operate the boats. In Gahirmatha, the team also captured mating turtles in offshore waters to tag them, using an ingenious triangular net that they devised (Shanker 2015).

The camps at Devi and Rushikulya both led to the birth of local community-based organisations. Recognising that they could continue working on sea turtles and generate funds for their work through both the forest department and other donors, many of these individuals formed their own NGOs. At Devi, the NGOs were led by Sovakar Behera or Tukku (Green Life Rural Association) and Bichitrananda Biswal or Bichi (Sea Turtle Action Programme), the latter barely a teenager when he first worked on sea turtles. In Rushikulya, the team that worked as Pandav's field assistants organized themselves as the Rushikulya Sea Turtle Protection Committee (RSTPC) led by the energetic Rabindranath Sahu (Rabi). In the 2000s, the team worked with Operation Kachhapa, a state-level programme to protect sea turtles, and as assistants for Basudev Tripathy and Suresh Kumar during their doctoral research at Rushikulya (Shanker 2009).

All these NGOS comprised a diverse crew of members, ranging from fishermen to agriculturalists and labourers from the coastal community who used their varied skill sets to organize themselves. Damburu in Rushikulya and Kalia in Gahirmatha, being traditional fishers, were irreplaceable in the value they added to field surveys with their knowledge of the sea and of turtles. Rabi and Tukku were essential in planning and coordinating local logistics and disseminating research results to the rest of the villages through their education programmes.

In the mid-2000s, several sea turtle biologists, principally Bivash Pandav, Basudev Tripathy, and one of the authors (KS), became interested in establishing a long-term monitoring programme for olive ridleys in Odisha. In 2005, with funding from the Marine Conservation Society, they trained the members of RSTPC and other members of the community in censusing mass nesting events using a standardized technique. In 2008, a long-term monitoring programme was set up by the Dakshin Foundation and the Indian Institute of Science at Rushikulya (Figure 10.1). Over the next few years, they started training forest department staff and engaged several field staff in the project, led by

*Figure 10.1* Field staff and researchers of the monitoring programme in Rushikulya, Odisha, after a night of counting turtles during an *arribada*

*Figure 10.2* Late Dhambru Behera, a senior field assistant in Odisha, explaining how the olive ridleys nesting at Rushikulya were his friends and would not harm anyone

*Photo: Muralidharan Manoharakrishnan*

Damburu Behera (Figure 10.2). The involvement of these staff in research projects over the previous decade played a significant role in the continued success of the programme, as they had already acquired the skills and experience for carrying out monitoring and research (Chandarana et al. 2017).

Each year at the beginning of the season, the mass nesting beach is mapped, as it is very dynamic and may shift in size and location from year to year. In addition, the programme also carries out surveys in the offshore waters to track the location and density of mating turtles. Some nests are collected and relocated to a hatchery where data on incubation temperatures are collected (Figure 10.3). After hatching, hatchlings that are found dead are collected and later dissected in the laboratory to determine their sex (Chandarana et al. 2017). In many reptiles including sea turtles, sex is determined by incubation temperature, with higher temperatures producing females, and lower temperatures, males (Bull 1980). Global phenomena such as climate change could result in temperature changes that alter population sex ratios and make them more vulnerable. Hence, the monitoring of temperatures and sex ratios seeks to determine the potential impact of climate change on these turtles.

*Figure 10.3* Olive ridley nests being relocated to a hatchery managed by Dakshin Foundation in Rushikulya, Odisha by field staff (Magata Behera, Bipro Behera, and Mahendra Nayak) from Purunabandha village

*Photo: Muralidharan Manoharakrishnan*

Mass nesting is sudden and unpredictable, the only pattern being that it seems to occur during the first or third quarter of the moon. Over the last decade, it has usually occurred in February or March in Rushikulya. Just before it begins, turtles can be seen closer to the shore near the nesting beach. With these signals, the monitoring team and the forest department gear themselves up to census the event. This is where years of training and practice counts – many of the researchers and field staff are adept at identifying which turtles to count. Since turtles often abandon nests and make multiple nesting attempts before they are successful, census staff must be able to identify turtles that are in the process of actually laying eggs – if not, the numbers are likely to be overestimated. Over the last decade, the estimates have ranged from 20,000 to over 300,000 turtles. In 2013, over 59,000 turtles were estimated to have nested on one of the nights (Chandarana et al. 2017).

The role of communities in monitoring sea turtles in Odisha has impacts far beyond the data that they help collect. Barring the historical trade of green turtles in the Gulf of Mannar (Jones and Fernando 1973) and a massive trade in olive ridley turtles in northern Odisha and West Bengal in the 1970s following the introduction of mechanization (Moll et al. 1983), there has not been organized harvesting of adult sea turtles in India. However, many communities along the coast have consumed sea turtle eggs (Tripathy and Choudhury 2007). Other than that, communities have largely been indifferent to sea turtles.

Over time though, increased conservation measures and restrictions on fishing and access to beaches has churned up a series of conflicts in locations such as Odisha and Lakshadweep. In Odisha, over 10,000 dead turtles have washed ashore each year since the 1990s, after accidentally getting caught and drowning in fishing nets, mostly trawls (Wright and Mohanty 2002). This led to many conservation campaigns which demonized fishing and created a deep rift between fishing communities and both the Forest Department, who are tasked with protecting turtles, and conservationists, who push for more stringent conservation measures. Even the fishing laws, which were originally passed to protect artisanal fishing in the 1980s, became a bone of contention, as they were primarily being invoked to protect turtles rather than fishers (Shanker 2015).

Even with growing resentment towards sea turtles and deep mistrust of conservationists, the engagement of local community members in activities associated with studying sea turtle behaviour and conservation has been the one factor that has had positive impacts on relationships with researchers/conservationists and attitudes towards sea turtles and conservation. The employment opportunities available with conservation groups and potential benefits of tourism are seen as pathways for alternate livelihood models. With proper planning and detailed scoping exercises, a sustainable model of community-based conservation can be developed for the region.

## Tracking turtles in the Andamans

It has been known for over a century that leatherback turtles are the ocean's great wanderers (Spotila and Tomillo 2015). Even though all sea turtles migrate large distances, leatherback turtles have been known to make spectacular trans-oceanic journeys, including to very cold waters in search of their jellyfish prey. In the Pacific, they have been tracked from beaches in West Papua, Indonesia to feeding grounds off the coast of north-western USA. Other turtles tagged at nearby beaches in Papua New Guinea went into the southern Pacific to the coast of South America. In the Atlantic, they have been tracked from the Caribbean to Canada. This means that very different threats at distant locations can affect these populations.

In the Andamans, much of our knowledge of leatherback turtles comes from Satish Bhaskar's seminal surveys starting in the 1970s and extending into the 1990s (Kar and Bhaskar 1982; Namboothri et al. 2012). Bhaskar, a drop-out from the Indian Institute of Technology, Chennai, began working on sea turtles with support from the Madras Snake Park and then the Madras Crocodile Bank. A self-taught sea turtle biologist, he surveyed much of the Indian coast and most of its offshore islands, including the remote

Nicobar islands during the late 1970s, and found significant leatherback nesting there. He returned to Great Nicobar Island in the early 1990s along with other researchers and established that Galathea on the south-eastern coast of Great Nicobar was a major nesting beach for leatherback turtles. In 2000, a monitoring programme was initiated at the site by the Andaman and Nicobar Environmental Team (ANET). Led by Harry Andrews and several other researchers, the backbone of the programme was its Karen field assistants.

The Karen are of Burmese origin and settled in the Andamans in the 1930s and 1940s, mostly in Middle Andamans near Mayabunder. They became involved in ecological research and surveys in the late 1970s and 1980s, including those by Bhaskar, and during visits by Romulus and Zai Whitaker, founders of the Madras Snake Park and the Madras Crocodile Bank. Whitaker is widely known as the 'snake man' of India and is well regarded for his work on crocodiles. Several members of the community such as Saw Pa-Aung (Uncle Pa-Aung) and Allen Vaughn played a key role in the initial years of surveying the islands. Later, Saw John helped coordinate the newly formed Andaman and Nicobar Environmental Team, and slowly more Karen staff began to work with the organisation. In addition, the research programme also attracted members of the 'Ranchi' community (settlers from Bihar) from Rutland Island and other islands.

For several years, the ANET team monitored the nesting beach at Galathea each night during the nesting season between November and March. For the first couple of years, all turtles that were encountered were tagged with Passive Internal Transponders (PIT) tags. Over 300 leatherback turtles were tagged during this period, which provided information on how often they nested during nesting seasons. As a part of long-term programmes, these tags can also provide information about remigration intervals, that is how many years turtles take between breeding. One of us (KS) spent the 2001–02 season at the camp (Andrews et al. 2006). Saw Agu, a Karen field assistant, who was then posted there, had remarkable field and data recording skills and could easily manage without any supervision.

In 2004 however, the Indian Ocean tsunami struck the islands and most of the nesting beaches in the Nicobars were completely destroyed. Saw Agu was present when the tsunami hit Galathea, and survived a 13-day ordeal adrift in the bay (Chandi 2009). The researcher at the sea turtle camp and several other visiting researchers lost their lives. Following this event, the research programme was closed down. In subsequent years, surveys indicated that some of the beaches had started forming again, but it was not possible to re-initiate the programme due to severe logistic constraints.

After conducting surveys of beaches in Little Andaman Islands in 2007, the Indian Institute of Science, ANET, and Dakshin Foundation initiated a monitoring programme at South Bay beach in 2008, and at West Bay in 2011. These beaches had far lower levels of nesting than the beaches on Great Nicobar Island, but could potentially be used as index beaches to monitor changes in the population. As always, the Karen team played a key role in the enterprise. The 2007 survey which led to the initiation of the programme was conducted with the help of Saw Pa-Aung as the boatman and Saw Agu. The camp was set up with two older Karen, Saw Bernie (Uncle Bernie) as boatman, and Saw Pamwein, as well as several younger staff notably Saw Thesorow – who would lead the camp for several years (Figure 10.4). In fact, in the early period, there was no field researcher present at the camp and the Karen team would manage all the data collection by themselves.

*Figure 10.4* Saw Thesorow and Saw John recording the clutch size of a leatherback nest in West Bay, Little Andaman Island

*Photo: Adhith Swaminathan*

The West Bay camp is completely isolated, a four-hour dinghy boat ride away from Hut Bay, the main settlement in Little Andamans. There is no cell phone signal anywhere nearby, the closest point possibly being the top of the lighthouse at South Bay, about two hours away by boat. The campsite is idyllic, by the side of a freshwater stream, with a few tents and shelters crafted by the Karen team. Apart from the supplies brought from Hut Bay, the Karen go out and catch a variety of fish, which makes life far more interesting during the two to three months that the team spend there, both for the cuisine and for something to do.

Since 2011, the monitoring camps have been coordinated by researcher Adhith Swaminathan and have expanded to include satellite telemetry of the leatherback turtles (Figure 10.5). Ten transmitters were deployed on leatherbacks. Of the first eight turtles, six travelled southeast, several along the coast of Sumatra, with a couple passing Cocos (Keeling) Islands, and one reaching the coast of Western Australia. Only one travelled southwest, towards the Chagos Archipelago and Maldives. While there appeared to be a pattern, the last two turtles tagged in 2014 both travelled southwest, reaching

*Figure 10.5* Saw Coloumbus, Saw John, Saw Thesorow, and Adhith Swaminathan attaching a satellite transmitter on a leatherback in the Andamans

*Photo: Kartik Shanker*

Mozambique and Madagascar on the east coast of Africa (Swaminathan et al. 2019).

Few of the surveys over the last three decades in the Andaman Islands for sea turtles and crocodiles would have been possible, or at least as rewarding, without the Karen collaboration. The monitoring camps at both Galathea and Little Andaman Island have depended on them running the camp and captaining the boats. Their knowledge of nature and seafaring has not only been critical for the logistics of surveys and monitoring, but enabled them to become keen observers of the biology of the fauna on which they have helped carry out research. In recent years, they have become involved with in-water projects, some of them learning scuba-diving, and have assisted with a range of other projects that researchers from various institutions carry out at ANET.

## Monitoring by networks

The above case studies involve monitoring led by researchers with the assistance of people from local communities and are focused on two index beaches. It is also still valuable to learn more about what is occurring to sea turtle populations over a larger geographical area. It may not be possible for a single or even several research institutions to cover the entire coast. Fortunately, there is a great deal of public interest in sea turtle populations. Over 70 countries globally have sea turtles that nest or forage along their coastlines and over half of these have some kind of non-profit or voluntary organization working towards sea turtle conservation. These organisations collect data with varying degrees of detail, which are valuable for creating long-term datasets.

In India, sea turtle conservation programmes started on the Chennai coast in the 1970s and were initiated by volunteers and non-governmental organisations such as the Madras Snake Park and Madras Crocodile Bank (Shanker 2003). From the late 1980s onwards, many NGOs have formed to address sea turtle conservation issues or have incorporated sea turtles into their conservation agenda. Today, there are at least one or two NGOs in each of the mainland's coastal states that have activities related to sea turtles, ranging from monitoring nesting beaches, protecting nests in-situ, protecting nests in hatcheries, turtle-related tourism and/or education, and outreach about sea turtles and coastal and marine conservation (Shanker 2015).

While the groups are all involved in sea turtle conservation, their motivation and aspirations are quite different. Theeram, a group comprised of fishermen in northern Kerala, became involved in the activity due to their discovery of the threats to marine turtles through a newspaper article but soon expanded to other aspects of environmental conservation such as coastal protection and mangrove afforestation. The Vishakhapatnam Society for Protection and Care of Animals is focused on animal rights issues. Others such as the TREE Foundation in Tamil Nadu, Sahyadri Nisarga

Mitra (SNM) in Maharashtra and Prakruti Nature Club in Gujarat seek to involve local communities in conservation.

The Students Sea Turtle Conservation Network (SSTCN) in Chennai (formed in 1988) patrols about 10–15 km of coast to protect turtle nests from poaching and depredation and relocates them to a hatchery. Citizens and students are engaged in the patrol walks at night during the nesting season (January to March) and in releasing hatchlings, which helps to engage society and build awareness of threats to sea turtles and marine habitats. Apart from the SSTCN, there are other smaller student groups that have formed over the years.

Many groups collect data on nesting abundances along the coast as a part of their ongoing activities. SNM in Maharashtra and SSTCN in Chennai frequently publish data on nesting and mortality in the Indian Ocean Turtle Newsletter (www.iotn.org). Other groups present their work at the meetings of the Turtle Action Group, a national network of these organisations. However, the challenge lies in gathering information in a systematic and organized fashion so that data can be pooled and analysed. Since the principal mandates of the groups are not research, this has not been easy to coordinate. Despite this, the fact that so many groups and individuals are contributing their time to monitoring increases not only the scale of the effort but also their independent motivation means that the provision of data to a collective effort is subsidized.

## Turtles and tourism

One other strategy that has the potential to include the community in a meaningful way and is rapidly gaining popularity is ecotourism. Ecotourism can provide employment, aid in community development, and inculcate pride in local efforts, while supporting conservation (Kale et al. 2016). Globally, sea turtle ecotourism initiatives are widespread in Australia, Sri Lanka, Costa Rica, Brazil, and India (Tisdell and Wilson 2001a, 2001b; Tisdell and Wilson 2005; Jacobson and Lopez 1994; Marcovaldi and Marcovaldi 1999; Kale et al. 2016).

Ideally, ecotourism caters to niche tourists who are more responsible in their activities and generate minimal impacts (Meletis and Harrison 2010). Ecotourism can create awareness about conservation, aid in revenue generation, and also bring about community development, participation and empowerment (Stem et al. 2003). Flagship species such as sea turtles can go beyond protecting other lesser-known species to also protecting socially disadvantaged communities (Shanker and Kutty 2005).

Sea turtles are considered especially attractive for ecotourism (Campbell 2003). In fact, sea turtles may be among the most popular animals on the planet for nature-based tourism (Senko et al. 2011). Tourism is so strongly linked to turtles in some locations that a decline in turtle populations can result in a corresponding loss in tourism revenue (Meletis and Harrison 2010). Hence, it has been argued that if turtles become a source of income for the community through tourism, the community will provide protection of the species, as they now hold some value. This is one way of striving for sustainability of both conservation and the tourism programme (Okazaki 2008; Stem et al. 2003).

Additionally, the skillset the community develops in the process can be used in other industries as well, also contributing to empowerment (Ashley and Roe 1998). For a long-term solution, mutually dependent, self-sustaining partnerships that strive for the overall development of the community, beyond just providing employment, are required. The involvement of the community in the decision-making process is critical for their acceptance and taking ownership of the venture (Ashley and Roe 1998). Inequity in the distribution of benefits to members of the community must be avoided, as this can incite differences, create new divides and exacerbate existing ones, leading to increased social stratification (Sproule 1996).

Sea turtle ecotourism initiatives have been attempted in Odisha (Chandarana et al. 2017), Maharashtra (Kale et al. 2016), and Goa (Dakshin Foundation 2020). Their success has depended on the involvement of local community members, assistance from other stakeholders, and adequate compensation for their efforts. In Velas, Maharashtra, community efforts have been successful in creating a functional ecotourism model developed in accordance with local needs and benefits. On the other hand, in Odisha and Goa, despite favourable factors such as larger tourist numbers and infrastructure, ecotourism efforts have so far enjoyed limited success due to a lack of enthusiasm at the community level and insufficient incentives to continue the projects.

## Conclusion

Citizen-led monitoring programmes can expand both the spatial scope of projects, as well as increase the number of data points, and help create valuable time-series data. One can also view the involvement of a wide range of civil society in ecological research as having a larger goal, one that goes beyond the narrow confines of data.

A unique by-product of research projects carried out along the coast is that the inclusion and curiosity of a few local participants as field staff in turtle monitoring programs has led them to becoming local advocates of conservation in the region. A better understanding of sea turtle biology and behaviour from their observations perhaps inculcated a local sense of pride in protecting the animals and their habitats. The arrival of external entities such as researchers can provide new forms of engagement and interesting perspectives about the turtles the locals have historically shared their spaces with.

Many environmental educators have also determined that the key to developing environmental values lies not in book-based learning (about say pollution, climate change, or habitat loss) but in developing a sensitivity to the environment. This is best brought about by experiential learning and observation. Hence, citizen science and community monitoring programmes offer an opportunity for students and civil society members to get hands-on experience – through observing birds, walking nature trails, collecting sea turtle eggs for hatcheries, or releasing hatchlings into the sea.

For communities, long-term monitoring programmes offer an opportunity to engage with science, which is viewed as a modern method of understanding and communicating about nature. Since these communities often have to engage with society and government, with regard to the resources they depend on or conservation issues in the areas in which they live, such programmes allow them to gain familiarity with the current language of negotiation (i.e. scientific data). For example, community groups involved in sea turtle conservation can provide forest departments with data on nesting population sizes. Similarly, fishing communities can collect data on catch which may help them in negotiating management measures with the fisheries department.

In the long run, such projects have multiple goals. They provide the data that are required for conservation strategies and management plans. Involving the local community expands the scope and the long-term sustainability of these projects. They also expand the constituency for the conservation of these species by garnering the support of the communities. Many recent campaigns for the conservation of turtles and their nesting beaches have involved or have even been led by local conservation groups. Finally, there may be wider livelihood opportunities for local communities beyond working on research projects in the form of eco-tourism, which provide an even greater impetus for conservation. Thus, community monitoring can be an effective long-term strategy to foster best practices, by building connections to nature and providing alternate livelihoods in conservation and research.

## References

Andrews, H. V., Krishnan, S., & Biswas, P. (2006). Distribution and status of marine turtles in the Andaman and Nicobar Islands. In K. Shanker & B. C. Choudhury (Eds.), *Marine turtles of the Indian subcontinent* (pp. 33–57). Hyderabad: Universities Press.

Ashley, C., & Roe, D. (1998). *Enhancing community involvement in wildlife tourism: Issues and challenges, wildlife and development series No. 11*. International Institute for Environment and Development. https://pubs.iied.org/7784iied

Balazs, G. H., & Chaloupka, M. (2004). Thirty-year recovery trend in the once depleted Hawaiian green sea turtle stock. *Biological Conservation*, *117*(5), 491–498. DOI:10.1016/j.biocon.2003.08.008

Bjorndal, K. A. (Ed.). (1995). *Biology and conservation of sea turtles*. Smithsonian Institution Press. doi:10.1017/S0025315400034020

Bull, J. J. (1980). Sex determination in reptiles. *The Quarterly Review of Biology*, *55*(1), 3–21.

Bustard, H. R. (1976). World's largest sea turtle rookery. *Tigerpasper*, *3*(3), 25.

Campbell, L. M. (2003). Contemporary culture, use, and conservation of sea turtles. In P. L. Lutz, J. A. Musick, & J. Wyneken (Eds.), *The biology of sea turtles, volume II* (pp. 307–338). London: CRC Press.

Carr, A. (1967). *So excellent a fishe: A natural history of sea turtles*. New York: American Museum of Natural History, Natural History Press.

Carroll, S. P., Hendry, A. P., Reznick, D. N., & Fox, C. W. (2007). Evolution on ecological time-scales. *Functional Ecology*, *21*(3), 387–393. https://doi.org/10.1111/j.1365-2435.2007.01289.x

Chandarana, R., Manoharakrishnan, M., & Shanker, K. (2017). *CMPA Technical Series No. 7: Long-term monitoring and community-based conservation of Olive Ridley turtles in Odisha*. Indo-German Biodiversity Programme, GIZ-India. www.indo-germanbiodiversity.com/pdf/publication/publication09-12-2017-1512808441.pdf

Chandi, M. (2009). Surviving the tsunami. *Current Conservation*, *3*(1), 16–19. www.currentconservation.org/surviving-the-tsunami/

Dakshin Foundation. (2020). *Final report: Galgibaga Turtle Conservation Reserve*. Bangalore: Dakshin Foundation.

Dickinson, J. L., Zuckerberg, B., & Bonter, D. N. (2010). Citizen science as an ecological research tool: Challenges and benefits. *Annual Review of Ecology, Evolution, and Systematics*, *41*, 149–172. https://doi.org/10.1146/annurev-ecolsys-102209-144636

Gates, C. E., Valverde, R. A., Mo, C. L., Chaves, A. C., Ballesteros, J., & Peskin, J. (1996). Estimating arribada size using a modified instantaneous count procedure. *Journal of Agricultural, Biological, and Environmental Statistics*, *1*(3), 275–287. https://doi.org/10.2307/1400519

Hamann, M., Godfrey, M. H., Seminoff, J. A., Arthur, K., Barata, P. C. R., Bjorndal, K. A., Bolten, A. B., Broderick, A. C., Campbell, L. M., Carreras, C., Casale, P., Chaloupka, M., Chan, S. K. F., Coyne, M. S., Crowder, L. B., Diez, C. E., Dutton, P. H., Epperly, S. P., FitzSimmons, N. N., . . . Godley, B. J. (2010). Global research priorities for sea turtles: Informing management and conservation in the 21st century. *Endangered Species Research*, *11*, 245–269. doi: 10.3354/esr00279

Jacobson, S. K., & Lopez, A. F. (1994). Biological impacts of ecotourism: Tourists and nesting turtles in Tortuguero National Park, Costa Rica. *Wildlife Society Bulletin*, 22, 414–419.

Jeffers, V. F., & Godley, B. J. (2016). Satellite tracking in sea turtles: How do we find our way to the conservation dividends? *Biological Conservation*, *199*, 172–184. https://doi.org/10.1016/j.biocon.2016.04.032

Jones, S., & Fernando, A. B. (1973). Present status of the turtle fishery in the Gulf of Mannar and the Palk Bay. *Proceedings of the Symposium on Living Resources of the Seas around India*, 712–715.

Kale, N., Muralidharan, M., & Shanker, K. (2016). The olive currency: A comparative account of community based ecotourism ventures in Western India. *Indian Ocean Turtle Newsletter*, *23*, 2–6. www.iotn.org/iotn23-02-the-olive-currency-a-comparative-account-of-community-based-ecotourism-ventures-in-western-india/

Kar, C. S. (1982). Discovery of second mass nesting ground of the Pacific Olive Ridley sea turtles in Orissa, India. *Tigerpaper*, *1*, 5–7.

Kar, C. S., & Bhaskar, S. (1982). Status of sea turtles in the eastern Indian Ocean. In K. A. Bjorndal (Ed.), *Biology and conservation of sea turtles* (pp. 365–372). Washington, DC: Smithsonian Institution Press.

Lutz, P. L., Musick, J. A., & Wyneken, J. (Eds.). (2002). *The biology of sea turtles, volume II*. London: CRC Press. https://doi.org/10.1201/9781420040807

Marcovaldi, M. A., & Marcovaldi, G. G. (1999). Marine turtles of Brazil: The history and structure of Projeto TAMAR-IBAMA. *Biological Conservation*, *91*(1), 35–41. https://doi.org/10.1016/S0006-3207(99)00043-9

Meletis, Z. A., & Harrison, E. C. (2010). Tourists and turtles: Searching for a balance in Tortuguero, Costa Rica. *Conservation and Society*, *8*(1), 26–43. DOI: 10.4103/0972–4923.62678

Mirtl, M., Borer, E. T., Djukic, I., Forsius, M., Haubold, H., Hugo, W., Jourdan, J., Lindenmayer, D., McDowell, W. H., Muraoka, H., Orenstein, D. E., Pauw, J. C., Peterseil, J., Shibata, H., Wohner, C., Yu, X., & Haase, P. (2018). Genesis, goals and achievements of long-term ecological research at the global scale: A critical review of ILTER and future directions. *Science of the Total Environment*, *626*, 1439–1462. doi: 10.1016/j.scitotenv.2017.12.001

Moll, E. O., Bhaskar, S., & Vijaya, J. (1983). Update on the Olive Ridley on the east coast of India. *Marine Turtle Newsletter*, *25*, 2–4.

Namboothri, N., Swaminathan, A., & Shanker, K. (2012). A compilation of data from Satish Bhaskar's sea turtle surveys of the Andaman and Nicobar Islands. *Indian Ocean Turtle Newsletter*, *16*, 4–13.

Okazaki, E. (2008). A community-based tourism model: Its conception and use. *Journal of Sustainable Tourism*, *16*(5), 511–529. DOI:10.1080/09669580802159594

Pandav, B., Choudhury, B. C., & Kar, C. S. (1994). Discovery of a new sea turtle rookery in Orissa. *Marine Turtle Newsletter*, *67*, 15–16.

Plotkin, P. T. (Ed.). (2007). *Biology and conservation of ridley sea turtles*. Baltimore: Johns Hopkins University Press.

Rees, A. F., Alfaro-Shigueto, J., Barata, P. C. R., Bjorndal, K. A., Bolten, A. B., Bourjea, J., Broderick, A. C., Campbell, L. M., Cardona, L., Carreras, C., Casale, P., Ceriani, S. A., Dutton, P. H., Eguchi, T., Formia, A., Fuentes, M. M. P. B., Fuller, W. J., Girondot, M., Godfrey, M. H., . . . Godley, B. J. (2016). Are we working towards global research priorities for management and conservation of sea turtles? *Endangered Species Research*, *31*, 337–382. https://doi.org/10.3354/esr00801

Robinson, N. J., Anders, D. A., Bachoo, S. A., Harris, L. I., Hughes, G. R., Kotzke, D., Maduray, S. E., Mccue, S. T., Meyer, M. I., Oosthuizen, H., Paladino, F.V., & Luschi, P. (2018). Satellite telemetry of leatherback and loggerhead sea turtles on the southeast African coastline. *Indian Ocean Turtle Newsletter*, *28*, 3–7. www.iotn.org/wp-content/uploads/2019/04/28-02-Satellite-tracking-of-leatherback-and-loggerhead-sea-turtles-on-the-southeast-African-coastline.pdf

Senko, J., Schneller, A. J., Solis, J., Ollervides, F., & Nichols, W. J. (2011). People helping turtles, turtles helping people: Understanding resident attitudes towards sea turtle conservation and opportunities for enhanced community participation in Bahia Magdalena, Mexico. *Ocean and Coastal Management*, *54*(2), 148–157. DOI:10.1016/j.ocecoaman.2010.10.030

Shanker, K. (2003). Thirty years of sea turtle conservation on the Madras coast: A review. *Kachhapa*, *8*, 16–19. www.seaturtlesofindia.org/wp-content/uploads/2017/04/Shanker-K.-2003.-Thirty-years-of-sea-turtle-conservation-on-the-Madras-coast-A-review.-Kachhapa-8–16–19.pdf

Shanker, K. (2009). Turtle boys. *Current Conservation, 3*(1), 23–25.

Shanker, K. (2015). *From soup to superstar: The story of sea turtle conservation along the Indian coast*. New Delhi, India: HarperCollins Litmus.

Shanker, K., & Andrews, H. (2006). *Monitoring and networking for sea turtle conservation in India: A UNEP CMS project report*. Mamallapuram, India: Centre for Herpetology/Madras Crocodile Bank Trust.

Shanker, K., & Kutty, R. (2005). Sailing the flagship fantastic: Myth and reality of sea turtle conservation in India. *Maritime Studies*, *3*(2) and *4*(1), 213–240.

Shanker, K., Pandav, B., & Choudhury, B. C. (2004). An assessment of the olive ridley turtle (*Lepidochelys olivacea*) nesting population in Orissa, India. *Biological Conservation*, *115*(1), 149–160. DOI:10.1016/S0006–3207(03)00104–6

Shimada, T., Jones, R., Limpus, C., Groom, R., & Hamann, M. (2016). Long-term and seasonal patterns of sea turtle home ranges in warm coastal foraging habitats: Implications for conservation. *Marine Ecology Progress Series*, *562*, 163–179. https://doi.org/10.3354/meps11972

Spotila, J. R., & Tomillo, P. S. (Eds.). (2015). *The leatherback turtle: Biology and conservation*. Baltimore: Johns Hopkins University Press.

Sproule, K. W. (1996). Community-based ecotourism development: Identifying partners in the process. In J. A. Miller & E. Malek-Zadeh (Eds.), *The ecotourism equation: Measuring the impacts* (pp. 233–250). New Haven, CT: Yale University RIS Publishing.

Stem, C. J., Lassoie, J. P., Lee, D. R., Deshler, D. D., & Schelhas, J. W. (2003). Community participation in ecotourism benefits: The link to conservation practices and perspectives. *Society & Natural Resources*, *16*(5), 387–413. https://doi.org/10.1080/08941920309177

Swaminathan, A., Namboothri, N., & Shanker, K. (2019). Tracking leatherback turtles from Little Andaman Island. *Indian Ocean Turtle Newsletter*, *29*, 8–10. www.iotn.org/wp-content/uploads/2019/04/29-03-Tracking-leatherback-turtles-from-Little-Andaman-Island.pdf

Tisdell, C., & Wilson, C. (2001a). Sea turtles as a non-consumptive tourism resource especially in Australia. *Tourism Management*, *22*(3), 279–288. DOI: 10.1016/s0261–5177(00)00059–5

Tisdell, C., & Wilson, C. (2001b). Wildlife-based tourism and increased support for nature conservation financially and otherwise: Evidence from sea turtle ecotourism at Mon Repos. *Tourism Economics*, *7*(3), 233–249. DOI:10.5367/000000001101297847

Tisdell, C., & Wilson, C. (2005). Does tourism contribute to sea turtle conservation? Is the flagship status of turtles advantageous? *Maritime Studies*, *3*(2) and *4*(1), 145–167. https://marecentre.nl/mast/documents/Mast-2004p.145-168.pdf

Tripathy, B., & Choudhury, B. C. (2007). A review of sea turtle exploitation with special reference to Orissa, Andhra Pradesh and Lakshadweep Islands, India. *Indian Journal of Traditional Knowledge*, *6*(2), 285–291. http://nopr.niscair.res.in/bitstream/123456789/922/1/IJTK%206(2)%20(2007)%20285-291.pdf

Troëng, S., & Rankin, E. (2005). Long-term conservation efforts contribute to positive green turtle *Chelonia mydas* nesting trend at Tortuguero, Costa Rica. *Biological Conservation*, *121*(1), 111–116. DOI:10.1016/j.biocon.2004.04.014

Wright, B., & Mohanty, B. (2002). Olive ridley mortality in gillnets in Orissa. *Kachhapa*, 6, 18.

# 11 Bringing reptiles into the conservation sphere

## A personal account

*Zai Whitaker*

## Introduction

As we pan across the scope and history of biodiversity conservation and sustainable use in India, it is interesting to zoom in on the induction of reptiles into conservation, a process which also brought in a new element of sustainable resource use. Today, the terms Snake Park, Serpentarium, 'Croc Bank', and snake rescuer among others, are commonly understood jargon. Their genesis lies in the story of an American herpetologist, Romulus (Rom) Whitaker, who arrived in Mumbai, then Bombay, in 1967 and stayed on to become an Indian citizen and acquire the title 'Snakeman of India'.

A decade earlier, in Grade 10 at a school in South India, Rom had written to one of his snake-handler heroes, William Haast of the Miami Serpentarium, and received an encouraging reply inviting him to visit if he ever found himself in Miami. Five years later, post (unfinished) college and looking for work on ships and a free passage to India to join his mother and siblings, he heard that Florida was a good possibility, as donated wheat was being transported from there to India under the PL480 food subsidy scheme. Additionally, he would be able to meet his hero. As it turned out, Haast was looking for a staff replacement, so it became a two-year visit, after which Rom was drafted into the US Army for service during the Vietnam War. Then, he finally travelled to India and by this time he had the best hands-on training in snake handling, venom extraction, and engagement with visitors – the three tenets of the Miami Serpentarium.

## The Madras Snake Park

The Miami Serpentarium prototype was transferred with Rom to India with modifications when he started the Madras Snake Park in 1969. Madras, now Chennai, was the perfect place for a snake-related project because of the presence of the Irular, who had been catching snakes for the skin industry for at least three generations. Staggering numbers are quoted for the annual export of skins – 10 million per year at its peak in the 1960s – and the Irular were probably the main suppliers. A unique example of a community whose livelihood was

DOI: 10.4324/9781003343493-15

based on sustainable forest use, yet involved in one of the most unsustainable wildlife industries ever.

At the Snake Park, the Irular played a major role in supplying snakes as well as in their care and husbandry. The site of the park, in a village some 35 kilometres south of the city, was practically made-to-order, with an abandoned construction site easily converted to several snake 'pits' that only required thatch roofing and a few more details. At 25 paise a ticket, the few visitors were able to see for the first time the common snakes of the area, alive and up-close. There were talks for visitors on both the harmless and venomous snakes, and the proper treatment of snakebite and demonstrations of venom extraction. Every year, thousands of people die of snakebites – the figure today is at 60,000 with many more left handicapped – leading to deeply entrenched superstitions about snakes and snakebites that the Snake Park team learnt about first-hand. When the park moved to the Guindy Deer Park in 1971 thanks to a Rs. 5,000 grant from the World Wide Fund for Nature (WWF), antivenom treatment awareness became a focal area.

The move to this unique city forest was enabled by the then Chief Conservator of Forests Mr. Bhoja Shetty. Mr. Shetty's decision turned out to be a fortuitous one for reptile conservation and for the Guindy Deer Park itself. When the then Prime Minister Indira Gandhi visited and was told about the threats to wildlife and forests from poaching and construction, she took quick action to make Guindy Deer Park/Sanctuary a national park. The 'centre pit' within the Park had a hundred or so common snakes including rat snakes, trinkets, striped keelbacks, and cobras and was a great attraction, although later abandoned due to husbandry challenges such as feeding and social domination by the larger species. There were hourly talks in English, Tamil, and Hindi about the importance of snakes in the environment, identifying the common ones, and the correct treatment of snakebite. A strong conservation and research element was added along the way and the Park's initiatives were extended beyond snakes to all reptiles.

## Running a Snake Park

I came into the picture around this time, marrying Rom in 1974, and making the transition from a bird-oriented home to a reptile-centred one. Over the next five years, the Snake Park's contributions included field surveys of endangered and endemic reptiles, snake rescues in the city, the first sea turtle hatchery, a volunteer program, and networking with reptile facilities in many countries. I started a quarterly reptile newsletter, *Hamadryad*, which carried articles and notes on the Park's field work which brought urgent issues and needs to public and official attention. Its publication continued until 2016, by which time it had become a serious herpetological journal with a hefty international board of editors. Forest Department officials from Tamil Nadu, Gujarat, and elsewhere were trained and assisted in marsh crocodile/mugger egg collection and hatching programs. The Snake Park also became the southern regional headquarters for WWF-India, and an early attempt at a uniform had a logo with a cobra on one side and a panda on the other.

Our range of conservation activities was made possible and sustained not by money, which was in short supply, but by a growing stream of visitors, and a team of passionate young people, many of them college dropouts, who were willing to work gratis for the cause of conservation. Several of them went on to play pivotal roles in the national conservation movement. Because of their commitment, Rom was able to spend much time in the field including a series of camps in Silent Valley, which lay the foundation of what was to become the first big environmental battle fought (and won) on the subcontinent. The Snake Park as well as these surveys and studies were supported by the income from the one million annual visitors. Administrative and secretarial costs were kept to a minimum and people joined the Park knowing that the remuneration would be more in the line of personal satisfaction rather than a bank balance. One of the gharial survey trips to the Chambal ended in bankruptcy and a 1,000 km hitchhike on a truck to Madras with a driver who felt sorry for the white 'Britisher' who had fallen on bad times.

The introduction of the *Wildlife (Protection) Act, 1972* brought with it several challenges, as it outlawed the use of several species hunted by the Irular for food. Additionally, large tracts of the evergreen coastal scrublands that were the hunting grounds and home of this hunter-gatherer community were now inaccessible because of the rapid and radial expansion of Madras city. In 1976, came the ban on the snakeskin industry, and therefore, the loss of the Irular's one constant and dependable source of income. Their abject economic situation and the acute shortage of snake venom for the production of antivenom were a perfect dovetail and laid the foundation of the Irula Snake Catchers' Industrial Cooperative Society (ISCICS), with the help of voluntary contributions of time and energy. It was registered in 1978 and based at the Snake Park. In 1984, with the Park's activities curtailed by confused ideas and actions by the group of trustees, we decided to focus our energies on the newly established Crocodile Bank. Our trusted team moved with us, as did the venom centre.

With donations in funds and equipment from Oxfam, the British High Commission, the Australian Embassy, and family and friends, ISCICS became and remains the only legal supplier of snake venom to laboratories that produce the polyvalent antivenom effective for the four common medically important snakes in India. These are popularly known as the Big Four – cobra, krait, Russell's viper, and saw-scaled viper. It is often cited as the perfect example of a 'sustainable' biodiversity project, where resources are utilized on a sustained yield basis without depleting wild populations (Table 11.1). Experts in the field receive advice on technical issues connected with the extraction process, storage, and laboratory management from platforms such as the Global Snakebite Initiative.

*Table 11.1* Snakes caught by ISCICS members and the venom yield (2018–19)

| *Name* | *Number* | *Venom produced* |
|---|---|---|
| Spectacled Cobra (*Naja naja*) | 2,000 snakes | 848 grams |
| Common Krait (*Bungarus caeruleus*) | 1,500 snakes | *61 grams* |
| Russell's Viper (*Daboia russelii*) | 2,000 snakes | *628 grams* |
| Saw scaled viper (*Echis carinatus*) | 2,800 snakes | *20 grams* |

Under licenses issued by the State Forest Department, Irular capture snakes from August to March, avoiding the breeding season for the harvested species. The snakes, which are housed in terracotta pots, are released after three extractions over the duration of a month. Today, ISCICS supplies to the six manufacturers of antivenom in the country, who all follow the same production process. Antivenom is made by immunizing horses with a cocktail of the four venoms. Thanks to their robust immune systems, the animals produce powerful antibodies that can bind to the snake venom components. Antivenoms are obtained by harvesting and then purifying the antibodies from plasma produced by the donor animal. This treatment works by boosting our immune response after a snakebite, enabling our own immune defences to eliminate these toxins.

As with the Snake Park, visitors buy tickets and can watch venom extraction, hear a talk about snakebite and its treatment, and interact with the Irular snake experts – amazing people who have now become internationally known through publicity in the media. Yet despite this recognition, it has been a challenging task to build leadership and decision-making systems within the Irular community, which has been side-lined and downtrodden for generations, losing its collective self-respect and confidence.

While there are some Irular women who catch snakes, it is mostly a male domain. Instead, women participate in rodent and termite hunts and gather plants, roots, and tubers for food and medicine. I felt it was important to create a forum for the economic and social empowerment of Irular women, with income generation from afforestation, a medicinal plant nursery, and a healing centre. The Irula Tribal Women's Welfare Society (ITWWS) was registered in 1986 and based in the village of Echur, where the first (and disastrous) afforestation project was undertaken.

The Wastelands Development Board had been recently constituted and ITWWS received a substantial grant to plant multi-layered sustainable forests on village lands including hardwoods, fast-growing border species, and fodder. It was amazing to watch the Irular's knowledge of seeds and seed gathering, germination and planting, and tending seedlings. Within three years, when it was beginning to look like a promising economic proposition, the agreement and understanding with the panchayat crumbled. Then on New Year's night of 1990, a gang of hoodlums armed with scythes and knives invaded and took over the property. Our staff escaped with their lives and a few possessions.

*Figure 11.1* The plant nursery of ITWWS

Source: *Rom Whitaker*

Shankar Ranganathan, a family friend and conservationist, heard about this debacle and financed the purchase of land in Thandari village in Thiruporur taluk, which borders the Chengalpattu Reserve Forest. Womankind Worldwide supported the construction of a centre in 1990–91 and remained a strong partner for many years. Today, its nursery, healing centre, and education programmes make a perfect trio for the use and development of traditional skills and knowledge (Figure 11.1). The healing centre, managed by traditional *vaidyars*, has had notable success with the treatment of leg swelling from filariasis, high blood sugar, and snakebite related morbidity.

ITWWS' social programs, funded by Womankind Worldwide, Kindernothilfe, NESA-Oxfam, and Action Aid among others, are run by village coalitions or *sangams*. The achievements include regularisation of land ownership deeds, drastic reduction in school dropout rates, legal support for rape and other crimes against women, health and hygiene initiatives including drinking water, and strengthening traditional skills and knowledge within the community, especially of Irular women. Visitation is much lower than at ISCICS, but the fact that the society has been extant and active for over three decades indicates the viability of this model. Today, ITWWS *sangam* leaders are able to meet police and administrative officials, push for widow benefits for its members, and participate in state and national Adivasi meetings and workshops. All leadership positions, including of the president, vice-president and directors, are held by Irular women.

## Madras Crocodile Bank

The Madras Crocodile Bank was registered as a Trust in 1976, having found a group of people willing to support its launch, with the objectives of captive breeding and conservation of the Indian crocodiles, as well as research. Two decades later, this was amended to include the conservation of all reptiles, including an international crocodile gene pool of the 23 species found around the world, and the tagline 'Centre for Herpetology' added to the Crocodile Bank's name. Starting small with just one crocodile pen and located on the tourist route to the famous Mamallapuram temples, it was self-sufficient from the beginning and grew dynamically as funds became more available, mostly from individual donors who financed a particular study, or breeding pen, or laboratory (Figure 11.2).

Today its visitation is 4.5 lakhs, with a total of 30 staff members. In addition to the visitors, camps, and programmes, presentations are organised for special groups such as students, teachers, and doctors on reptile

*Figure 11.2* The newly completed gharial enclosure at the Croc Bank gets its first inhabitant

Source: *The Madras Crocodile Bank Trust*

conservation and biology, their importance in the environment, and strategies for decreasing human–reptile conflict including the prevention and treatment of snakebite. Similar to the Snake Park, the Crocodile Bank's core campus has supported a radial network of conservation-related activities and I will focus on three of these that have been particularly effective – field work, field stations, and snakebite mitigation.

### *Field work*

The Snake Park's field work continued at the Crocodile Bank (Croc Bank), as the conservation team had made the move with us. The field work and conservation follow-up has been wide-ranging in scale. The geography of study sites includes over 15 states and many species from the four reptile groups – snakes, chelonians (turtles and tortoises), crocodilians, and lizards. Two outstanding former members of the team were Satish Bhaskar and J. Vijaya, both of whom focused on chelonians: Satish on sea turtles and Vijaya on tortoises and freshwater turtles.

Low-cost transport (often their two legs), outdoor camping (no tents!), and a close connection with local communities were the hallmarks of their studies, which contributed much new information on species range and distribution, natural history, and human use. Lobbying and publicity brought with it conservation actions, such as the closure of the sea turtle blood drinkers' market in Tuticorin, the captive breeding of endangered species such as the Travancore tortoise and red-crowned roof turtle, and rewilding the mugger and gharial in many locations. Other species in this program included pythons, northern river terrapins, and red-crowned roof turtles.

Apart from one salt-water croc nest collected in the Andamans, the seed animals for the Croc Bank were all mugger, hatched from eggs collected in wild habitats in Tamil Nadu, with permission from the Forest Department. These locations included large reservoirs such as the Mettur and Amaravathi dams, remote riverine habitats in the upper reaches of the Moyar, and small rainwater ponds adjoining rice fields. Camping on the banks of these varied habitats, with Irular teammates pointing out tracks and slide marks invisible to us, we were able to learn and substantially record information about mugger natural history.

### *Field station: Agumbe*

The field work increasingly highlighted the need for research and conservation hubs situated in biological hotspots. Three have been created and are successful centres of education, research, and conservation. The first one, the Agumbe Rainforest Research Station (ARRS), is located in the rainforest stretch of Agumbe in the central Western Ghats, the Shimoga district of Karnataka state. Rom first visited Agumbe in 1972 at the suggestion of his friend Kenneth Anderson and caught a pair of king cobras using only

*Figure 11.3* The king cobra rescuers of Agumbe (Ajay Giri, Rom, and Kumar)
Source: *Zai Whitaker*

his sleeping bag and a stick, an incident that was widely reported in the national press. This was the beginning of a fascination with Agumbe, a place of unreal rainfall, magnificent forests, and friendly people who revere the same snake he does, the king cobra. Other iconic herpetofauna in the area include draco (flying lizard), pit vipers, and several species of endangered amphibians.

Over the next three decades, there were many more visits and attempts at creating the field station with this iconic species at its centre (Figure 11.3). This became possible in 2005 when Rom was able to buy 10 acres of land in the centre of this rainforest, thanks to a legacy left by his mother, Doris Norden. Receiving the Whitley Award for Nature as well as later additional funding from the Whitley Fund and then the Rolex Award enabled the construction of visitor and volunteer accommodations, a library, a laboratory, and an office.

The Forest Department allowed the team to start the first telemetry study on the king cobra at this site. Dr. Matt Goode from the University of Arizona at Tucson came to advise on the implant surgery and for the past decade, the two phases of the telemetry study have been ably conducted by local community persons and volunteers from India and abroad. Telemetry is a wonderful tool and enables observations and insights that would not otherwise be possible to collect on these secretive snakes. Today, ARRS activities include a free snake rescue service and are supported through visitors, volunteers, and some continuing grants from the Deshpande Foundation and the King Cobra Conservancy based in the United States. Recent grants include a jeep for the rescue project from the Wildlife Conservation Trust and many generous individual donors.

### *Field station: The Andaman and Nicobar Islands*

The Andaman and Nicobar Environment Team (ANET), another Croc Bank field station, was started in 1989 by Rom, my father, and myself and followed the same pattern of surveys and collection trips, leading up to a conservation centre supported by family, friends, and conservation organisations (Figure 11.4). Its genesis was from a 1975 reptile collection trip for the Snake Park. The team comprised an Irular snake catcher called Annamalai and myself. The Andaman and Nicobar Islands were still the land of magic – uncontacted tribesmen living out stone-age cultures, high levels of plant and animal endemics (many still to be discovered), and a breath-taking tapestry of coral reefs, rain forest, pristine beaches, and mangrove forest.

*Figure 11.4* The Aka-Bea, ANET's first research boat

Source: Zai Whitaker

In tandem with the excitement of the 'discovery' of this paradisiacal ecosystem with its many reptile endemics was the anxiety of the burgeoning anti-environmental development plans that were already afoot and would come to fruition all too soon. ANET lobbied long and hard against these plans, suggesting viable alternatives which could sustain both the local communities and support carefully managed tourism. Rajiv and Sonia Gandhi's visit to the Snake Park gave us the opportunity to point out the problems facing the islands. As Prime Minister, he took some positive steps to stop the rampant deforestation.

ANET also carried out surveys on the archipelago's herpetofauna and assisted the government in biodiversity conservation, including planning and managing protected areas. As one of the oldest and most committed NGOs in the Islands, ANET played a pivotal humanitarian role after the 2004 tsunami devastated the region, including providing relief supplies for desperate victims. Today, although still active in the herpetological arena, ANET carries out a broad range of environmental work and follows the tried-and-tested model of conservation, research, and education. Its education and awareness-building work includes programmes and presentations for local and mainland schools, colleges, and other institutions on the amazing ecology of these Islands and the need to conserve it.

From the earliest surveys to the present day, conservation work in the Andamans and Nicobars would have been impossible without the help and partnership of the Karen community, many of whom are valuable field assistants with their excellent boat skills, forest knowledge, and generous character. Dakshin Foundation has been a partner at ANET for two decades, helping oversee the research, administration, and development. In 2019, it took over the organisation, which was an easy and productive transition, as the Foundation's objectives and goals align with those of the Croc Bank.

### *Field station: Garhaita*

2000 kilometres to the north of the Croc Bank on the mighty Chambal River lies Garhaita, the base of the Gharial Ecology Project, which is one of the most long-term and in-depth crocodile studies in the world. This has been possible because of the relentless work of Professor Jeffrey Lang, with whom the Croc Bank has been in partnership since the early 1980s. Using sophisticated data collection techniques including telemetry, the study tracks the home ranges and observes the habits of this fascinating and critically endangered crocodilian. Local inhabitants are being trained in field techniques and the conservation of the species. Interesting new data and information have emerged about the natural history of the gharial, such as the extensive riverine range of adults and the protection of multiple hatchling creches by adult males. Funded by the Prague Zoo as well as multiple individual donors and zoos in the United States, the Gharial Ecology Project works closely on the conservation of the species across its tri-state Chambal habitat, which spans Uttar Pradesh, Rajasthan, and Madhya Pradesh.

## Snakebite mitigation

The 40 years of close contact with the terrible and largely unnecessary realities of snakebite deaths (upwards of 60,000 annually) kick-started the Snake Conservation and Snakebite Mitigation project in 2016. Funded by USV Pharmaceuticals and others, the team explores and finds ways to address several aspects of the problem. The most immediate is to spread awareness about snakebite prevention. Hundreds of programmes focused on this angle were conducted in the six states with the highest snakebite deaths in the country – Jharkhand, Orissa, Bihar, Madhya Pradesh, Andhra Pradesh, and Tamil Nadu.

Another part of the project involves improving antivenom quality and including a wider geographical representation of venoms. Venom composition and resultant effects of bites vary regionally, but at present, only snakes from Tamil Nadu (mostly from one or two districts) are used, thus limiting the efficacy of the polyvalent antivenom. Discussions are underway with the government and other organisations such as the Indian Council for Medical Research, Medicins sans Frontieres, Madras Medical College, CMC Vellore and WHO, about this and other snakebite issues.

## Conclusion

I am often asked what I feel was the most important contribution of Rom's 50-year-and-counting conservation journey in his adopted country. To this trick question, I always answer that first, these projects and organisations created a new level of conservation leadership. Many of the 'flunkies' who chose to work with us rather than pursue more lucrative careers, now have the commitment and conviction to speak up against the many environmentally damaging decisions being made, at a time when the words 'sustainable development' sound increasingly oxymoronic. The other part to my answer is that it engineered the acceptance of reptiles as wildlife and as integral links in a healthy and productive environment. This journey included struggles with deep-rooted superstition, public and official apathy, and challenging bank balances, but ultimately has made conservation in India stronger.

# 12 Linking ecotourism and biodiversity conservation

## Lessons from India

*Kunal Sharma and Lokesh Kumar*

### Introduction

Travelling and exploration are essential elements of human nature as these experiences open our eyes to the unique wonders of our world while providing thrill and wisdom simultaneously. Tourism is an educational activity as well (Tomasi et al. 2020), for it confers new insight into the natural history, culture, and varied traditions of communities. It is an opportunity not only to discover joy in the exquisiteness of the destination but also to ensure that the splendour remains undisturbed and cherished by future generations. Tourism in its pure form is an invigorating mélange of sights, smells, tastes, and sounds of the world we live in.

While tourism itself is extremely old and well established in many societies (Butler 2015), in recent times, as means of transportation improved and paid vacations became popular, there was a rush to visit popular destinations. This form of travel was loosely termed as conventional tourism or mass tourism. It was the great era of discovery and virtually the entire planet became one huge tourist destination. The Americas, Europe, and Oceania emerged as popular destinations for globetrotting. Soon, however, this infrastructure-intensive form of tourism was found to have significant environmental, socio-economic, and cultural impacts. Mass tourism often spurred uneven development and amounted to high social and environmental costs, and so alternative sustainable tourism means had to be sought (Honey 2008).

The growing discord with mass tourism led to a clamour for a more authentic experience. Thus, in sync with growing environmental awareness in the 1980s (Bagri et al. 2009) grew a desire to travel responsibly, which we know now as 'ecotourism'. It sought an optimistic re-imagination of tourists as discerning people who are conscious of the ecological and economic costs of their travels to natural areas. The earliest definition of ecotourism was developed in the year 1983 by Hector Ceballos-Lascurain. Subsequently, The International Ecotourism Society in 2015 defined ecotourism as 'responsible travel to natural areas that conserves the environment, sustains the well-being of the local people and involves interpretation and education' (Mondino et al. 2018).

DOI: 10.4324/9781003343493-16

*Figure 12.1* Guests at Binsar Forest Retreat, a private ecolodge model for ecotourism in India

*Photo: Pallavi*

The growing discourse around ecotourism sought to meet the threefold criteria of facilitating environmental conservation, community participation, and promoting livelihood generation, and by the mid-1990s, ecotourism was hailed as the fastest growing sector of the tourism industry (Erdogan and Erdogan 2010). Through economically valuing biodiversity, ecotourism aspired to protect the natural areas where it was established from further degradation (Figure 12.1). In some cases, tourism actively complemented biodiversity conservation and even claimed credit for partly protecting vulnerable species (Larm et al. 2017; Steven et al. 2013).

However, the loss of biodiversity is a complex and common accusation against ecotourism (Michael Hall 2010). Even its most ardent supporter cannot fight off the criticism that ecotourism renders biodiversity vulnerable (Nash 2001) and can only claim to conserve the environment indirectly. Yet despite criticisms of the ecotourism movement, the ideologies prevailing in the past three decades have helped bring in an era of environmentally friendly tourism. With its potential for providing employment in remote regions, preserving large tracts of land for a better natural experience, and generally being a non-consumptive industry, ecotourism has a constantly evolving connection with the sustainable use of biodiversity. In fact, the incentive value for preserving natural environments is now acknowledged as the biggest benefit of ecotourism (Weaver 1998).

Yet since many of the destinations are located in ecologically, as well as culturally sensitive areas, the anticipated impacts might be more severe and irreversible than is understood (Weaver 2001). However, from a purely economic point of view, it is within the interest of tourism operators to preserve and sustainably use the natural resources from which they earn income or else the quality of the experience suffers. Thus, the key is to manage ecotourism's interaction with nature within a sustainable framework in order to continue protecting biodiversity, generating local livelihoods, and supporting private enterprise.

## Negative effects of ecotourism

Ecotourism ventures are frequently established in pristine and fragile ecosystems. Improper understanding of the principles of ecotourism and poor planning often results in a disproportionately large ecological footprint of ill-planned eco-ventures that run the risk of destroying the very resources upon which they depend. Some of the prominent impacts of poorly planned ecotourism include the following:

- The tourism industry can cause significant environmental damage in the form of natural habitat loss (Ozturk et al. 2015).
- Eco-ventures around national parks may go beyond the carrying capacity (Manning 2002).
- Facilities can lead to generation of solid waste and pollution (Sunlu 2003) in areas with little capacity to absorb them.
- Inordinate pressure on natural resources from excessive use of energy, ground water extraction, and disturbance to natural areas (Belsoy et al. 2012).
- Vehicle safaris and inappropriate visitor behaviour can disturb wildlife (Higginbottom and Buckley 2003).
- Light pollution from eco-lodges adjacent to forests can stress wildlife.

The issues regarding ecotourism extend to its definition as well, with long-standing ambiguity making it difficult to precisely define ecotourism. Ecotourism is often confused with other terms that people consider synonymous and interchangeable, such as sustainable tourism, nature tourism, wildlife tourism, adventure tourism, and responsible tourism. Over 85 published definitions of ecotourism exist broadly based on the notions of 'conservation', 'sustainability', 'education', and 'local benefit' in varying formulations (Fennel 2001), portraying the lack of a commonly accepted definition.

This ambiguity has led to accusations of 'green washing' (Abeyratne and Arachchi 2021) when some entrepreneurs freely use the term ecotourism for short-term financial profits without integrating sufficient sustainable practices into their operations. For example, permanent concrete buildings are often passed off as ecotourism ventures, in direct contrast to vernacular, truly environmentally conscious lodges. There is deep concern that ecotourism is being employed as a marketing tool, rather than a resource management one (Telfer and Sharpley 2015). In India, there is a twofold reason why 'green' marketing tactics continue to thrive. Firstly, there is absence of a stringent certification policy or a code of conduct for tourism facilities around protected areas which can monitor compliance and grade sustainable practices. Secondly, consumers' travel choices continue to be based on popularity, price mechanisms, access, and provision of facilities (Puri et al. 2018).

Ecotourism has immense potential to contribute to local livelihoods with its unique selling point of providing 'jobs at the doorstep'. Ecotourism and especially non-consumptive wildlife tourism (Curtin and Kragh 2014) provides unique opportunities to alleviate rural unemployment. However, very often ecotourism merely results in small, spasmodic cash gains for a local community. Most profits go to local elites, outside operators, and government agencies. Only a few individuals or families gain direct financial benefits from ecotourism, while others cannot find a way to share in these economic benefits because they lack capital or appropriate skills (Scheyvens 1999).

Solely profit-driven eco-ventures are likely to be damaging to the environment, undermining the efforts of more responsible competitors who practice low-impact tourism and already suffer from low visitation, especially during initial years of operation. This may compromise the quality of experience and conservation efforts. Therefore, people are calling for certifications of ecotourism ventures from competent authorities to fix the problems that arise from casual interpretations of ecotourism principles.

## Ecotourism in India

Unlike in several countries where it grew as a demand from enthusiasts, ecotourism in India has largely been a supply-driven activity. Traditionally, tourism in India was oriented towards popular destinations, pilgrimages, and local travel. However, when ecotourism grew internationally, the industry in India also grew, as people began choosing to travel to pristine destinations based on word-of-mouth advertising and around theme-based interests such as wildlife, birding, beaches, crafts, and culture.

Ecotourism expanded in the past three decades with the creation of experiences that matched the latent desire among prospective tourists to visit natural areas in India. The wide variety of choices complemented the sheer diversity of the nation – from Corbett, Bandhavgarh, Ranthambore, Kanha, and Periyar National Parks, which offered wildlife sighting opportunities, to the states of Goa and Kerala that offered a pristine beach experience, and destinations such as Rishikesh and Haridwar that promoted yoga. Among the various offshoots of ecotourism, it is nature-based tourism that has been growing rapidly in India (Karanth and Defries 2011).

Ecolodges gained prominence as an offshoot of the ecotourism industry. Either privately owned or under government control, the key attribute of ecolodges is that they meet the philosophy and principles of ecotourism (Hawkins et al. 1995) and have the potential to influence a guest's attitude towards nature. Ecolodges are usually located near protected areas, typically small in scale, often use environmentally friendly design, employ locals, and offer guests an educational experience that can develop into long-lasting ties with dedicated clientele. Though largely ignored in policy, ecolodge operators serve an important role in promoting ecotourism.

In other parts of the country, private entrepreneurs set up eco-ventures, which focus on providing activities for guests, such as bird-watching trips, safaris, and guided walks. Designed as a high-yielding product, the ecotourism supply model attracts a large number of international customers enticed by tales of the Indian wilderness. Occasional visits from high profile figures also help boost the fame of national parks such as Bandhavgarh and Ranthambore. In 2002, the Ministry of Tourism launched the Incredible India campaign as a major marketing strategy that helped develop India's tourism brand (Mishra and Sajnani 2020). Various theme-based tourist circuits were offered to the world – wildlife, culture, religious (e.g. Buddhist), beaches, architecture, local cuisine, etc.

Additionally, high profile travel agencies tapped into the international travel market with fixed itineraries and succeeded in bringing a significant number of tourists into India's national parks. While ecotourism grew at a relatively fast pace around protected areas, niche homestays also developed, especially in the mountainous regions of the country. Currently, the demand for ecotourism products is highly sought after by domestic visitors also.

## Central and state policy on ecotourism

Tourism has immense potential in India. In 2020, the travel and tourism industry's contribution to the GDP was US$121.9 billion – this is expected to reach US$512 billion by 2028. The industry's direct contribution to India's GDP is expected to record an annual growth rate of 10.35% between 2019 and 2028. Besides, in 2020, the Indian tourism sector accounted for 31.8 million jobs, which was 7.3% of the total employment in the country. By 2029, it is expected to account for about 53 million jobs (IBEF 2021). Ecotourism policy in India has been shrouded in dualism since the beginning. While, on the one hand, policymakers acknowledge its potential contribution to economic growth, tourism and especially ecotourism is often seen as disruptive to the conservation of fragile forest resources.

In 2002, after years of discussion and planning, the Ministry of Tourism and Culture, Government of India created a tourism policy designed to position tourism as an engine of economic growth for employment and poverty eradication in an environmentally sustainable manner. The past 20 years has not seen the policy being updated, although a draft national tourism policy is currently being discussed. Foreign tourist arrivals in 2002 were 2.4 million, in contrast to the several million international arrivals in 2019 (Bhutia 2022). Additionally, tourism is a state subject. Attempts are being made to place it in the concurrent list (i.e. both the centre and states can legislate on it), which is expected to provide much needed momentum to the sector (Tewari 2021).

However, it bears mentioning that central and state tourism schemes usually promote infrastructural development at the expense of creating an authentic ecotourism experience. This often involves maintenance of existing structures, providing for public amenities in mass tourism destinations or building new structures in wilderness areas – all in the name of ecotourism development. Few policies address community participation or reflect the other founding principles of ecotourism. In fact, for most government bodies, tourism is a minor department.

There are also inter-departmental tussles involved. A large majority of natural landscapes comes under the purview of the state Forest Department, to whom ecotourism has become an additional burden on top of routine forest management. In some areas however, motivated officials have created unique models of ecotourism, often involving community participation. Significantly, the Ministry of Environment, Forest and Climate Change in its National Wildlife Action Plan 2002–16 referred to 'Tourism in Protected Areas', recognising that regulated, low impact tourism has the potential to be a vital conservation tool in winning public support for wildlife conservation (EQUATION 2010). Nevertheless, lack of jurisdictional clarity with regard to ecotourism has constrained the effective implementation of ecotourism policies.

India largely lacks a dedicated structure for establishing state ecotourism boards. Only a few states have set up ecotourism boards, which are crucial in coordinating with other departments including Tourism, Forest, Rural Development, Skill Development and with the private sector to develop ecotourism destinations. States such as Kerala, Himachal Pradesh, Madhya Pradesh, Uttarakhand, Sikkim, and Karnataka took early leads in developing ecotourism-centric policies and implementation guidelines, with their main objectives being:

- Conservation of natural resources
- Livelihood opportunity to locals
- Education to tourists
- Quality of standard accommodations
- Stakeholder participation
- Public–private partnerships

However, despite this drive, there have been few successful models of ecotourism. The low success rates may be attributed to the following factors that call for remedial policy initiatives for making ecotourism more sustainable.

**Ownership:** Since protected areas come under the purview of forest departments, they must take a lead in ecotourism development, in association with other departments for joint implementation.

**Scarcity of Budget:** Forest departments lack funds for developing ecotourism sites. They should identify and use funds from other departments, as well as from private funding institutions.

**Zonation & Land Banks:** Land adjacent to protected areas is often unavailable for tourism. Planners should prepare land banks along with the local community.

**Ecotourism Planning:** An implementation plan based on policy recommendations is critical for the success of ecotourism projects.

**Institutional Mechanism:** Each state should have a dedicated Ecotourism Board or committee for the implementation of ecotourism policy.

**Carrying Capacity:** Ecotourism destinations should be planned in accordance with the carrying capacity of the area, as this will discourage degradation.

**Resource Mapping:** Maps can help identify core primary resources and feasible activities around protected areas to create the right ecotourism product mix.

**Integration with Central Government Initiatives:** Schemes like Swadesh Darshan with its eco-circuit, tribal, wildlife, Himalayan, and coastal circuits, and schemes implemented by the Ministry of Tourism should be incorporated into state-level ecotourism planning.

**Viability Gap Funding:** Since ecotourism projects operate in remote locations, private sector entrepreneurs may not find such ventures financially viable. However, to benefit from private sector expertise, social entrepreneurs should be encouraged with viability gap funding opportunities to make their projects sustainable.

**Green Washing:** Ecotourism Certification has to be issued by credible agencies to be empaneled under the Ministry of Tourism, who shall issue accreditation to the desired tourism ventures after thorough analysis of defined parameters. The process of issuing ecotourism certification needs to include an analysis of green washing in private projects by the issuing agency. This will check the misuse of jargon (like green, ecotourism, sustainable, zero carbon footprints etc.) to attract serious ecotourists to a destination.

**Exclusivity:** Certain tourism activities should be reserved for local communities to ensure equitable distribution of economic benefits, drawing on successful prototypes like the eco-development model of Periyar, Kerala.

## Models of ecotourism in India

In India, ecotourism models have evolved to work with diverse landscapes, community groups, and economic interests in developing site-specific projects. Central and state governments, local community groups, corporations, non-governmental organisations, private homestays, and small-sized entrepreneurs have all contributed to the development of this niche tourism sector. In the coming century, ecotourism initiatives are expected to contribute significantly to improving the health of the natural habitats where they are located and provide livelihood opportunities to local communities. The following section discusses ecotourism initiatives that have helped conserve biodiversity in the country and will hopefully shed light on future developments in this fledgling industry.

### *Government ecotourism model: Jungle Lodges and Resorts Limited, Karnataka*

Jungle Lodges and Resorts (JLR) is a wholly owned corporation of the Government of Karnataka. As most of its resorts are located near protected forests, the organisation has, over the past four decades, focused on creating motivated 'ambassadors of conservation'. Key policies adopted by the organisation include zero tolerance for loud music in the campus, effective waste management, and a strict adherence to safari ethics. JLR is credited with several innovations in the ecotourism industry, such as identifying new destinations, thereby reducing adverse impacts on the main tourism hotspots. It also promoted the idea of weekend getaways to off-beat locations in close proximity to major cities such as Bengaluru and Mysuru.

Additionally, JLR developed thriving ecolodges around themes that cater to different interest groups, ranging from a beach to a wildlife-based experience. JLR also partnered with professional adventure operators, thus improving safety standards in adventure activities. JLR created template schedules with integrated packages and fixed time slots for safaris, meals, treks, and other activities. This process of automation removed the need for excessive specialisation as the staff could multitask with ease. From its earliest days, the organisation adopted the principle of enhancing local livelihoods, making efforts to employ local villagers. Several staff have gone on to occupy high positions in the organisation and continue to do so today. Furthermore, JLR has made continuous efforts to share the lessons learned on ecotourism through regular trainings and associations with other state governments.

Although the JLR model is a work in progress, especially on issues related to maintaining an optimum competitive cost, effectively managing its increasing carbon footprint, improving waste disposal systems, managing limitations of a semi-skilled workforce, conducting proper orientations to guests on the ethics of ecotourism, and fostering more community connection, this government model of ecotourism has many important lessons to share with the rest of the country. JLR demonstrates that an effective allocation of government infrastructure to a credible organisation can open opportunities for travel to unexplored regions. JLR also has a long-standing model of integrated tour packages that allow for continuous engagement with guests. The organisation's strict adherence to state-mandated wildlife safari norms regulations contributes to reduced disturbance to wildlife. The organisation also supplements the efforts of the Forest Department by working to mitigate forest fires and serving as the 'eyes and ears' of the Department in tourism zones.

### *Private ecolodge model: Binsar Forest Retreat, Uttarakhand*

The Binsar Forest Retreat, a small family-run property located amid an oak and rhododendron forest in Binsar Wildlife Sanctuary can offer several enduring lessons to ecotourism lodges around the world. Sustainability is ingrained in the ethos of this unique model, as the owners are highly sensitive about their carbon trail. They continuously evolve practises to reduce their footprint and do not use generators in order to maintain the tranquil atmosphere of the surrounding forests (Figure 12.2).

The Retreat runs instead on a 2KW solar panel that generates the required energy. The team did away with most electric appliances such as refrigerators, geysers, room heaters, electric kettles, and blenders. Food is prepared using the age-old process of stone grinding or using a mortar and pestle. Since light and water are finite resources in the hills of Kumaon, guests are

*Figure 12.2* Team BFR at Binsar Forest Retreat
*Photo: Pallavi*

*Figure 12.3* Room interior at Binsar Forest Retreat

*Photo: Pallavi*

asked to be mindful of their usage. While drinking water comes from an underground spring in the forest through a gravity-based pipeline, rainwater is harvested for all other purposes. There are no showers in the bathrooms and buckets are provided instead (Figure 12.3).

The practice of waste management on the property goes beyond mere lip service with the understanding that the full responsibility for the waste produced is borne by the Retreat. Anything wrapped in plastic is not allowed on the property. Garbage is segregated and goes to the local *kabaddi wala* (recycler), who either reuses or recycles all waste. Items that cannot be recycled are disposed of in the government landfill. Wet waste gets composted. Black and grey water go into soak pits where they undergo slow decomposition. Organic cleaning agents such as a combination of soap nut and wood ash are used for washing dishes.

Conservation of the forest ecosystem and the preservation of its original state is a shared objective with the local villagers, the Retreat, and the Forest Department. A constant watch for forest fires is carried out and the Retreat reports any fire immediately to the Forest Department. The Retreat contributes men and materials to the Forest Department for putting out fires as soon as possible. Additionally, in order to reduce the spread of pine species that are very prone to fire, the hosts are planning to extensively plant native trees around the area.

*Figure 12.4* Nature walk at Binsar Forest Retreat
*Photo: Pallavi*

Hosts at the Retreat make a conscious effort to spread awareness among their guests about local flora and fauna. The idea is not only for guests to enjoy their stay but also to take back with them an awareness of conservation values. The hosts regularly accompany guests on forest walks to guide them to tread lightly in the ancient forest (Figure 12.4). With a focus on protecting the fragile ecosystem, the Retreat also grows native plants that attract birds, butterflies, and other insects rather than having manicured exotic gardens.

The Retreat supports the local economy by employing staff from nearby villages. Seasonal produce such as rice, millets, dry fruits, and walnuts are locally procured. To highlight the age-old cultural practices of the hills, three women from the local village come to the retreat once a week during the tourist season to de-husk rice, earning up to Rs. 1,200 each. Additionally, a family in a nearby village has taken on laundry management, with a washer and iron box set-up. With the Retreat's support, another family has started a dormitory to accommodate the drivers accompanying guests. In total, the small retreat of eight rooms financially supports about 18 local families including 11 staff, one dormitory owner, one laundry person, three village women for de-husking, and two men for local transport.

The Retreat also uses the services of local nature guides for birding and trail walks. The guides earn up to Rs. 1,000 for a full day's walk. Moreover, the hosts have recently initiated a village stay program where guests can experience village life and spend a night with local families. Though nascent, the community project has received positive feedback and provides a stable source of income for local families.

The Retreat's community outreach has contributed to reducing the exodus of villagers from the hills. Some staff members who had previously worked in cities such as Haldwani, Lucknow, and Delhi now choose to stay in the hills as a result of the better quality of life. The Binsar Forest Retreat showcases how a small property can adopt numerous actions to reduce its carbon footprint and negate the adverse effects of tourism. With a low carrying capacity, sensitive staff, effective waste management, clean energy systems, employment generation at the local level, and vigorous forest conservation, not only does the small operation uphold the principles of ecotourism but also has the potential to motivate others to replicate this sustainable ecolodge model from the hills of Uttarakhand.

### *Private home stay model: Raju's homestay, Himachal Pradesh*

Often located in pristine regions and with personal stake in preserving the local ecology, niche homestays have the potential to highlight the advantages of ecotourism. Raju Bharti's homestay is one such family-operated enterprise located deep in the heart of the Kullu District of Himachal Pradesh. Considered a pioneer of the homestay movement in the Western Himalayas, Bharti's homestay is situated along the banks of the pristine Tirthan River, which is the habitat of the prized rainbow and brown trout (Figure 12.5). This area is among the key entry points to the Great Himalayan National Park and designated by UNESCO in 2014 as a World Heritage Site of 'outstanding significance of biodiversity conservation'.

Raju Bharti initiated this project in the early 1990s when tourism was unheard of in this valley. Today, his example showcases how a small, family-run homestay can become a sustainable venture. Together with his wife Lata, who personally attends to and prepares food for the guests, he rented out two rooms at his traditional house and then added a few more rooms subsequently. Their sons Karan and Varun, their wives and grandchild help look after guests and manage bookings. Outside of family, the homestay provides employment to six people from nearby villages.

In the early days, only enthusiastic anglers or trekkers made the treacherous journey to this remote village. This ardent repeat clientele helped develop a unique identity for the homestay. Slowly, with the advent of the internet, the valley became popular and experienced a spurt in tourist infrastructure development in the form of homestays, cottages, camps, and resorts. However, Raju remained unfazed by this and has maintained the family-owned-and-operated ethos of his property.

*Figure 12.5* The grounds outside of Raju's Homestay
*Photo: Karan*

The family has remained steadfast in their conviction that tourism activities should be inherently sustainable. Hence, they uphold strict guidelines on the premises, especially when it comes to waste management. They discourage the use of single-use plastic and food waste is either given to the yard animals or turned into compost. Karan and Varun are closely associated with the Tirthan Conservation and Tourism Development Association, a group of young tourism professionals from the valley who are committed to protecting wildlife, educating locals about sustainable tourism, and ensuring effective waste management. Additionally, they plant various tree species including Himalayan cedar, apricot, chestnut, and oak every year during the monsoon season to increase the green cover and reduce erosion.

However, conservation in this region has not been an easy task. Some years ago, hydroelectric projects were proposed along the Tirthan River and its tributaries, which may have put an end to the rich biodiversity of this pristine valley. Raju, along with his father and a local NGO, filed a complaint in the High Court against these projects. The battle lasted for 5 years in Court, before the river and its people emerged victorious. Because of Raju's persistent efforts and of the people who stood by his beliefs, Tirthan is one of the few remaining freshwater Himalayan rivers not choked by hydroelectric projects in the name of development and green energy.

The family believes that increasing influx of tourists puts severe social and ecological pressure on natural destinations. Therefore, they have limited the number of rooms and continue to maintain an orchard on the remaining land. The expansive five-acre orchard produces a wide variety of fruits including cherries, plum, pear, apricot, peach, apples, and persimmon, which are used to make the juices, jams, and pickles. Dairy is sourced from their own cows and the dung is used as manure for vegetable and fruit gardens. Remaining steadfast to the concepts of ecotourism, vernacular construction techniques were used to build the additional cottages, using local construction materials including wood, slate, and stone.

The biggest fear of entrepreneurs like Raju is that of unplanned development and unprofessional operations in the pristine valley. Though he looks forward to receiving more tourists, he is convinced that ecotourism is ideally practised through small, family-owned properties. Large resorts or campsites may provide more employment and revenue in the short term but will eventually put too much strain upon the carrying capacity of the region and lead to the downfall of this sought-after tourist destination. Small entrepreneurs like him serve as role models for youths who want to start sustainable homestays in fragile mountain destinations.

### *NGO-led initiative: Last Wilderness Foundation and the Pardhis, Madhya Pradesh*

Forest-dependent Adivasi communities have a deeply symbiotic relationship with nature. Their knowledge of the landscape, subsistence-based hunting of wild animals, and collection of wild food sustain their sustainable, foraging lifestyle. Building upon the intrinsic knowledge of one such group, the Pardhis of Central India, a civil society organisation – Last Wilderness Foundation (LWF) began an initiative in 2016 for establishing an ecotourism-based livelihood model. Initially, the foundation engaged with youth on wildlife conservation activities. As the Pardhis are known for their intimate knowledge of the forest and unparalleled tracking skills, the group built on and redirected these abilities towards offering a guided wilderness experience for ecotourists (Figure 12.6).

LWF collaborated with Taj Safaris and the Panna Forest Department in Madhya Pradesh to introduce an initiative called 'Walk with the Pardhis' where they trained 25 Pardhi youths including four women on guiding techniques to enhance their existing knowledge of biodiversity. Once qualified,

*Figure 12.6* Waterfall landscape at LWF field site
*Photo: Last Wilderness Foundation*

the youth actively engaged with Taj Safaris (Figure 12.7). This initiative has also opened the doors for more youths to engage in activities such as cooking traditional food for visitors and preparing wildlife-themed handicrafts. Other resorts in the Panna and Kanha landscapes are keen on hiring these trained guides as naturalists.

Economic empowerment has helped many youths to convince their families to turn away from illegal activities such as poaching. Community members have also started exploring alternative livelihood options such as poultry farming. After learning of the ecotourism opportunities in Panna, Pardhi community members from other locations have voluntarily approached LWF to start similar initiatives in their areas. This is a huge win–win for the communities and for conservation. This collaborative effort offers hope for an effective forest protection system, while providing a dignified livelihood to often-stigmatized Adivasi communities.

The project also explores the role of ecotourism with respect to gender roles. Traditionally, apart from managing their homes, Pardhi women also take up the responsibility of earning a living for the family. Interestingly, women do

*Figure 12.7* Pardhi naturalists with guests of Taj Safaris on a guided walk
*Photo: Last Wilderness Foundation*

not see working as nature guides as an alien activity because it involves engaging with the forests and wildlife, traditionally an integral part of their lives. Employing women as guides is a big step forward in terms of the local ecotourism industry, as most traditional tourism jobs employ women largely for the stereotypical gendered roles of cooking, washing, and cleaning. Even in eco-lodges, women often carry a disproportionate burden of tiring chores without having any significant voice in the operation of such units. The LWF initiative in this regard brings in not only economic but also social empowerment while providing visibility and mobility to local Pardhi women (Figure 12.8).

### *Community-based ecotourism projects*

Based on the case studies from India such as the ones discussed earlier, it is evident that the success of an ecotourism project is based on developing sturdy institutional structures. These critical structures include dedicated community members, enduring entrepreneurship skills, and effective government departments, such as Forest and Tourism, who continue to play a major role in supporting the institutional framework of community-led

*Figure 12.8* Women at work threshing and hand-pounding

*Photo: Last Wilderness Foundation*

enterprises. In line with these, we see that community-based tourism projects such as the Himalayan Homestays in Ladakh and Lahaul & Spiti, the Mountain Shepherds Initiative in Uttarakhand, and the Manas Maozigendri Eco-tourism Society in Assam have benefited adjacent protected areas where they function and the local stakeholders that are involved. These projects educate tourists about the flora and fauna of the region and the unique culture of local communities (EQUATION 2010).

Many more unique models of community-based ecotourism have set the direction for future ecotourism ventures, including the Velas Olive Ridley Homestay model in Maharashtra, Eaglenest Sanctuary in Arunachal Pradesh, the Blue Yonder in Kerala, Ecosphere in Spiti, Himachal Pradesh, Help Tourism in Assam, and Snow Leopard Conservancy in Ladakh & Spiti, among others. The eco-development project of the Periyar Tiger Reserve (PTR) is a particularly important model in this regard. Widely discussed, this World Bank-funded model was initiated under the India Eco-Development Project during 1996–2004. The key features of the project which made it so successful are relevant to discuss here.

The first critical initiative of the Project was the formation of 72 Eco-Development Committees (EDC) and Self-help Groups (SHGs) which were trained to perform various primary and supporting roles related to ecotourism in the tiger reserve. These positions included naturalists, trekking guides, boatmen, bamboo rafting guides, camp operators, cooks, support staff, patrolling squads, LPG providers, and kitchen assistants, among others. The second initiative provided skill training and capacity building for each member of the community to prepare them for their professional duties. The third initiative designed successful ecotourism products such as tiger patrolling, bamboo rafting, boating at the Periyar Lake, border hiking, night patrolling, the Mangladevi trek, the Bamboo Grove Campsite, and sandalwood forest patrolling.

The project received worldwide acclaim and two EDCs even won accolades at the national level. The first comprised former *vayana* bark collectors, who had previously illegally harvested cinnamon bark, but then became guards responsible for conservation at Periyar and conducting a trekking expedition called the 'tiger patrol'. The second EDC was *Vasanthasena* which is an all-women group responsible for patrolling and conserving the remaining patches of sandalwood forests in the Periyar Tiger Reserve.

The success of PTR was a result of the synergy between the Forest and Tourism Department who jointly formed the Ecotourism Directorate. They shared the responsibilities of the conservation of the tiger reserve while simultaneously providing world class tourist-centric activities. Additionally, all responsibilities for the operational aspects of tourism activities were exclusively allocated to the community and EDCs, thus improving profitability while contributing to forest conservation. The Periyar model was a ground-breaking achievement for ecotourism ventures as it supported forest conservation while providing livelihood opportunities to the local community. Adherence to forest regulations, upholding the carrying capacity, and environmental education through a visitor interpretation centre were ingrained into the project.

While this is a highly feted model, it has not been widely adopted by other states. However, the project is replicable in other protected areas and has the potential to scale-up and address the issues of marginalised communities in fringe areas of forests. The model showcases that providing communities with exclusive rights to conduct tourism activities through the formation of user groups, with the support and synergy between the Forest and Tourism department, and continuous skill building, can contribute to ecotourism while ensuring biodiversity conservation.

Another unique community-based ecotourism project is the UNESCO funded Sikkim Himalayan Homestay Project. The project worked towards promoting, developing, and standardising the homestay network by focusing on village-based tourism at four places – Yuksom, Pastangla, Kewzing, and Dzongu. While the communities benefited from an alternate source of income in the form of low-impact tourism, the project helped showcase the rich natural and cultural heritage of the state of Sikkim to tourists. The project was a model in community mobilisation, with the formation of village-level tourism development committees for joint decision making and conflict resolution. A rotation system for booking was set up so that every homestay owner got an equal chance of hosting and earning income. The committees ensured that a portion of the booking amount was used for conservation activities to protect the rich biodiversity of the region.

With the right institutional mechanism in place, the Ecotourism and Conservation Society of Sikkim (ECOSS) became a project partner and contributed to skill training, product development, operations, general etiquette, and guest training. The project achieved success as the community-run homestays professionally managed activities such as village walks, birdwatching, nature trails, local cuisine and drinks, traditional healing practices, agro-tourism, rural immersion programs, crafts, festivals, and many more.

## Lessons from successful eco-ventures

While ecotourism in India still has a long way to go, well-run ventures show immense potential for conserving biodiversity especially around protected areas and easing undue pressure on crowded national parks. Active socio-economic involvement of local populations is a key component that separates ecotourism from other forms of tourism. Ecotourism can constructively play the dual role of providing alternate livelihood options to communities while promoting conservation. Based on the lessons realised through the above models, some possible future pathways for ecotourism in India are listed below:

- Community leadership is a vital component for ecotourism to thrive. As community members become better equipped to assume ownership, they will be incentivized to make effective decisions for conservation, social development, conflict resolution, and benefit sharing.
- Key government departments such as Forest, Tourism, Rural and Skill Development have the potential to play a critical role in the success of ecotourism, especially when collaborating with local communities.
- Ecotourism projects often perish due to lack of effective marketing. State tourism departments can play a major role along with conventional and online tourism aggregators to stimulate interest among prospective clientele in ecotourism ventures.
- The development of most eco-ventures is personally financed by individuals. However, local communities living adjacent to protected areas are usually marginalised groups and indigenous people, who cannot afford to contribute to the capital costs. Therefore, initial capital support for eco-ventures is vital and could be provided to local communities by state governments.
- Ecotourism is dependent on primary tourism resources. Hence, ecotourism planners should actively contribute to the planning phase. Participatory planning processes such as Appreciative Participatory Planning and Action (APPA) should be used jointly with communities to establish rights over their ownership in the project from the beginning.

## Conclusion

In order to ensure that ecotourism projects are effectively implemented and sustainable in the long run, planners should consider approaching the subject through the eyes of the primary stakeholders – at the individual and community level, as well as state and national actors. At the community and individual level, it is essential to support nascent projects in planning, capacity building, and product development with capital subsidies in the initial years of operation. Ensuring that communities have a vital stake in the ecotourism project as showcased by the success of the Periyar model, strengthens long-term community involvement in the project.

At the state level, both Forest and Tourism Departments should strive for an enhanced budget designated to identifying and undertaking destination development. The departments should put in place action plans, policies, and guidelines to develop ecotourism projects in states. Public–private partnership projects which ensure community leadership should be provided with viability gap funding to meet capital requirements. The central government could initiate a push towards undertaking more ecotourism projects through schemes like the Swadesh Darshan, in consultation with state governments.

Successful eco-ventures need to consistently adopt a three-pronged approach to improve their appeal to prospective customers. They need to improve their operational efficiency by being water, energy, and waste positive, they need to improve their brand appeal by promoting sustainable product choices, and they need to improve community stewardship in conserving the environment through effective partnerships.

## References

Abeyratne, S. N., & Arachchi, R. S. S. W. (2021). Ecotourism or green washing? A study on the link between green practices and behavioral intention of eco tourists. In A. Sharma & A. Hassan (Eds.), *Future of tourism in Asia* (pp. 51–63). New York: Springer. DOI: 10.1007/978-981-16-1669-3_4

Bagri, S., Gupta, B., & George, B. (2009). Environmental orientation and ecotourism awareness among pilgrims, adventure tourists, and leisure tourists. *Tourism*, *57*(1), 55–68.

Belsoy, J., Korir, J., & Yego, J. (2012). Environmental impacts of tourism in protected areas. *Journal of Environment and Earth Science*, *2*(10), 64–73.

Bhutia, P. D. (2022, March 14). India needs a new tourism policy – Now. *Skift*. https://skift.com/2022/03/14/india-needs-a-new-tourism-policy-now/

Butler, R. (2015). The evolution of tourism and tourism research. *Tourism Recreation Research*, *40*(1), 16–27. https://doi.org/10.1080/02508281.2015.1007632

Curtin, S., & Kragh, G. (2014). Wildlife tourism: Reconnecting people with nature. *Human Dimensions of Wildlife*, *19*(6), 545–554. https://doi.org/10.1080/10871209.2014.921957

EQUATION. (2010, October 15). Tourism in and around Pas – A paradigm shift needed. *Equitable Tourism Options*. http://equitabletourism.org/blog-post/tourism-and-around-pas-paradigm-shift-needed.

Erdogan, I., & Erdogan, N. (2010). A critical evaluation of ecotourism. In TODEG (Ed.), *Ecotourism in forest ecosystems workshop & TODEG in its tenth year* (pp. 65–81). TODEG (Foresters' Association of Turkey Ecotourism Group). www.researchgate.net/publication/235913942_A_Critical_Evaluation_of_Ecotourism

Fennell, D. (2001). A content analysis of ecotourism definitions. *Current Issues in Tourism*, *4*(5), 403–421. https://doi.org/10.1080/13683500108667896

Hawkins, D. E., Epler Wood, M., & Bittman, S. (Eds.). (1995). *The ecolodge sourcebook*. The Ecotourism Society.

Higginbottom, K., & Buckley, R. C. (2003). *Terrestrial wildlife viewing in Australia*. Gold Coast, Queensland: CRC for Sustainable Tourism.

Honey, M. (2008). *Ecotourism and sustainable development: Who owns paradise?* (2nd ed.). Island Press.

India Brand Equity Foundation. (n.d.). *Indian tourism and hospitality industry analysis*. www.ibef.org/industry/indian-tourism-and-hospitality-industry-analysis-presentation

Karanth, K. K., & DeFries, R. (2011). Nature-based tourism in Indian protected areas: New challenges for park management. *Conservation Letters*, *4*(2), 137–149. https://doi.org/10.1111/j.1755-263X.2010.00154.x

Larm, M., Elmhagen, B., Granquist, S., Brundin, E., & Angerbjörn, A. (2017). The role of wildlife tourism in conservation of endangered species: Implications of safari tourism for conservation of the Arctic fox in Sweden. *Human Dimensions of Wildlife*, *23*(3), 1–16. DOI:10.1080/10871209.2017.1414336

Manning, R. E. (2002). How much is too much? Carrying capacity of national parks and protected areas. In A. Arnberger, C. Brandenburg, & A. Muhar (Eds.), *Conference proceedings: Monitoring and management of visitor flows in recreational and protected areas* (pp. 306–313). Institute for Landscape Architecture and Landscape Management, Bodenkultur University. https://mmv.boku.ac.at/downloads/mmv1-proceedings.pdf

Michael Hall, C. (2010). Tourism and biodiversity: More significant than climate change? *Journal of Heritage Tourism*, *5*(4), 253–266. DOI: 10.1080/1743873X.2010.517843

Mishra, R., & Sajnani, M. (2020). Is incredible India campaign a reason for growth in tourism? A case study of Madhya Pradesh. *Amity Research Journal of Tourism, Aviation and Hospitality*, 2(2), 28–35.

Mondino, E., & Beery, T. (2018). Ecotourism as a learning tool for sustainable development. The case of Monviso Transboundary Biosphere Reserve, Italy. *Journal of Ecotourism*, *18*(2), 107–121. https://doi.org/10.1080/14724049.2018.1462371

Nash, J. (2001, April 1). *Eco-tourism: Encouraging conservation or adding to exploitation?* Population Reference Bureau. www.prb.org/resources/eco-tourism-encouraging-conservation-or-adding-to-exploitation/

Ozturk, A. B., Ozer, O., & Çaliskan, U. (2015). The relationship between local residents' perceptions of tourism and their happiness: A case of Kusadasi, Turkey. *Tourism Review*, *70*(3), 232–242.

Puri, M., Karanth, K. K., & Thapa, B. (2019). Trends and pathways for ecotourism research in India. *Journal of Ecotourism*, *18*(2), 122–141. DOI: 10.1080/14724049.2018.1474885

Scheyvens, R. (1999). Ecotourism and the empowerment of local communities. *Tourism Management*, *20*(2), 245–249. https://doi.org/10.1016/S0261-5177(98)00069-7

Steven, R., Castley, J. G., & Buckley, R. (2013). Tourism revenue as a conservation tool for threatened birds in protected areas. *PLoS ONE*, *8*(5), e62598. https://doi.org/10.1371/journal.pone.0062598

Sunlu, U. (2003). Environmental impacts of tourism. In D. Camarda, & L. Grassini (Eds.), *Options Méditerranéennes, Série A., Séminaires Méditerranéens, Number 57: Local resources and global trades: Environments and agriculture in the Mediterranean region* (pp. 263–270). Centre International de Hautes Etudes Agronomiques Méditerranéennes.

T3 News Network. (2022, February 1). Mixed reaction from the industry to budget 2022–23. *Travel Trends Today*. www.traveltrendstoday.in/news/india-tourism/item/10666-mixed-reaction-from-the-industry-to-budget-2022-23

Telfer, D. J., & Sharpley, R. (2015). *Tourism and development in the developing world* (2nd ed.) New York: Routledge.

Tewari, S. (2021, August 3). Ministry bats for tourism's inclusion in concurrent list. *Mint*. www.livemint.com/companies/news/tourism-likely-to-be-included-in-concurrent-list-11627997824209.html.

Tomasi, S., Paviotti, G., & Cavicchi, A. (2020). Educational tourism and local development: The role of universities. *Sustainability*, *12*(17), 6766. https://doi.org/10.3390/su12176766

Weaver, D. B. (1998). *Ecotourism in the less developed world*. CAB International.

Weaver, D. B. (2001). Ecotourism as mass tourism: Contradiction or reality? *Cornell Hotel and Restaurant Administration Quarterly*, *42*(2), 104–112. https://doi.org/10.1177/0010880401422010

# 13 Sacred groves of Central India

## Beyond the botany and the ecology

*Madhu Ramnath*

### Introduction

> **sak-** To sanctify. 1. Suffixed form **sak-ro-*, a. SACRED, SACRISTAN, SEXTON; CONSECRATE, EXECRATE, from Latin *sacer*, holy, sacred, dedicated
>
> (Watkins 2011)

Much has been written about sacred groves in India. Mostly, these writings have dealt with the richness of plant species in these pockets of green, which more often than not provide a safe haven for animals and birds in the area. These groves have been recognised as 'conserved areas' of the communities who live in the proximity and are extolled as a traditional method of conservation. The intention in this chapter is not to overlook the importance, in terms of conservation and ecology, of sacred groves but to look at their social and religious roles in the lives of the communities who nurture such spaces. Unfortunately, the earlier coherence of many communities has not remained as before – various influences, from panchayat politics to education, roads, and market forces have all played their part. Quite appropriately, perhaps even obviously, the state of the grove reflects that of the community.

The material and ideas for this chapter have come from visits to and inquiries about a hundred sacred groves in erstwhile Bastar and Koraput districts in Chhattisgarh and Odisha, respectively. The communities are of the Bhatra, Durwa, Koitoor, Desiya, and other Adivasi peoples. The author has visited most of the groves personally; a few were visited by members of the Legal and Environmental Action Foundation, based in Bastar, and the information passed on. The broad genera of information collected and used for this chapter encompass:

- the size of the grove
- the unusual plants found in the grove

DOI: 10.4324/9781003343493-17

- the threats faced by the grove
- the role of the grove in community life, and
- stories about the origins of the grove, the deities in these shrines and the relationship between people and the deities.

This last point also hints at movements and migrations of these communities, suggesting reasons, for instance, of why the Durwa people live among the Bhatra in certain parts of Bastar. The attempt through the consideration of these 'non-ecological' aspects of sacred groves is to understand less tangible but equally important and real matters for the communities that nurture groves and live with them.

## The sacred grove of Koleng

Several years ago, I walked into a five-star hotel in Delhi with a tribal friend from Bastar. The lobby was artfully decorated in the modern ethnic style with various statues of Hindu gods – the large bronze dancing Nataraja and the smooth stone Ganesh were imposing presences. My friend observed them and asked, *Kya inka bhi puja hota hai?* (Do they make offerings to these too?) And I told him that this was 'art' and that the figures were not there for a religious purpose. His response was that any stone will remain a stone until it is given the power and the respect that we want it to have. He added that a stone can be given sufficient power to make it a god. Light an incense stick at the foot of this Ganesh, my friend said, half-jokingly, and see how things change. I did not do that, but the incident remained with me and came back to life when I visited the sacred grove of Koleng in Bastar.

This grove in central Bastar is about eight acres in its spread but getting smaller, as some parts of it have been nibbled at by various peoples of influence. (The latest of these detrimental forces that has reduced the grove is the Central Reserve Police Force that now has a base on a rocky outcrop within the grove). A botanical inventory of the grove revealed that 70% of the trees were > 1 m in girth – the minimum girth measured was 60 cm, the maximum being 4.7 m for a mango tree of which the people in the village were extremely proud. It was customary to not climb the mango trees in the grove and during the time of their fruiting the trees were smeared with lime powder. Any person seen with white dust on his body was an obvious thief who was then taken to task.

The Koleng village has more than 10 hamlets and, like any village, continues to spread. When a home grows with a new bride or a child, which inevitably leads to another hearth, new homes and shelters are constructed, and the village grows. Only one hamlet, the one next to the sacred grove, called the par pārā, remains as it was in size. Nobody here built a new house, which means that families did not split up when a new bride came in, and the proverbial mother-in-law, daughter-in-law conflicts were resolved within their homes. There was really no choice for the people of this pārā, as a new home would mean that they had to leave the pārā altogether and settle somewhere in the outskirts of the village (away from their kinsmen) because the grove limited all expansions. They could not possibly clear a section of the grove nearest to them to build a house. The grove exerts an influence on the people who lived closest to it. The atmosphere in this part of the village was tangibly different, with the large grove fringing a part of it, the taller of the trees strikingly visible from most homes. It would not be an exaggeration to add that, perhaps, because of the large extended families living under one roof in these homes, an overall sense of coherence prevailed.

The grove itself is set in the middle of a sāl (*Shorea robusta*) forest and comprised of a dense stand of 47 tree species, predominated by *Mangifera indica* and *Schleichera oleosa* (with 78 and 71 individuals, respectively), and with small trees, climbers, and other plants comprising another 67 species. (A botanical puzzle that has occupied me is the absence of sāl within the grove – one answer I received for this is the below-ground mycorrhizal interactions). In one part of the grove with abundant giant lianas, a separate shrine is dedicated to the Earth Goddess (*Welka guddi*), where the 'new fire' ceremony happens each year. All the fires in the village are put out once each year and a new fire is coaxed out of two pieces of bamboo. This new fire is then carried out from the grove and taken to all the homes, remaining in the village for the year.

## The sacred grove in community life

Sifting through the information given by the priests and some of the elders about the groves in their villages, one comes across a lot of similarities in the roles that these sacred spaces play across many communities. The maintenance of village peace, meaning the well-being of the people and the cattle, is apparently the most important role that groves seem to have. This is how the grove, and the deity residing there respond to the offerings and the prayers of the community. In most cases, this extends to the belief that non-compliance with the appropriate behaviour results in the kind of retribution that the community wants to avoid. In addition, in many villages, meetings to decide upon important matters that affect the community are held in the grove. The community feels that with the deity as a witness to the discussions, people will speak the truth and good decisions will be made. Other ways in which the grove contributes to overall community well-being are expressed as follows:

- To maintain peace in the village and community
- To prevent the spread of illnesses
- To assure full granaries
- To keep wild animals away from the village and to protect people who guard their crops at night, etc.

There are also specific requests brought to the deity in the grove and these include cures from skin infection, diarrhoea, and smallpox, for sufficient rain, to keep cattle healthy, and to bless childless couples with offspring. There is a grove where a person who strayed away from the community could be reinstated into it with appropriate ceremonies and rituals. Another grove is for the well-being of one specific community. Some groves serve as places where pre-sowing and post-harvest ceremonies happen. And there are several groves where religious fairs are organized annually, bi-annually, or less frequently. Such fairs are large events, drawing people from distant villages, with *sirās* (shamans) congregating for special dances and sacrifices. The reputation of villages that host such gatherings is a notch or two higher than villages that do not.

## The sacred space

The mention of sacred groves usually conjures up images of dense forest patches. But that may not be true. The geography of sacred spaces in central India varies from a thick sāl forest to a large bare rock with a single banyan tree. Other spaces include a clump of a few trees, a mango orchard of old trees (common in the Malkangiri district of Odisha), a hillock, and a shrine in the middle of rice fields. The smallest of these groves is almost bereft of any tree, the shrine being the last vestige of what was once a grove. The largest is a wonderful mixed forest of a little over a hundred acres (in Sandh Karmari village, Bastar district, Chhattisgarh), with giant trees and lianas, and rare terrestrial orchids growing in its shady spots.

Though the forest where these groves are located has a vast number of plant species, the groves themselves are known for specific plants that are rare in the vicinity. People make journeys to these sites to see rare plants, such as the cane in the Durkabeda grove in Bastar district, the gulvel (*Tinospora cordifolia*) of Sandh Karmari, or the giant gudi-pal (*Capparis horrida*) of Koleng. Other plants mentioned include bhuinim, tikhur, sirandi, bhuikurwa, indrajal, dev-badan, kukur-shendi, kakai, jili, jadi, chandan, bhelwa, bhatri, kudai, baila-gondri, vajradanti, siyadi, baheda, harra, amla, gangosiyali, and menda. Though these plants may be found in the surrounding forest tracts, many of them are of great medicinal value and/or rare.

Menda is *Litsea glutinosa*, which has all but disappeared from much of central India (and ironically, this is an ingredient in incense sticks lit in temples and shrines all over Asia). Kukur-shendi is *Rauwolfia serpentina*, exploited by pharmaceutical companies and now rarely encountered in the forest. Tikhur, *Curcuma leucorhiza*, which yields a form of arrowroot, is seldom seen in the wild but cultivated in homesteads. Many of the other plants have medicinal value (bhuinim, *Swertia chirayita*, kudai, *Holarhenna antidysentrica*), however over the years, they have been totally eradicated from the wild – the last remnants of these species are found in sacred spaces, guarded by a socio-religious fence of faith. But can such fences withstand the strength and schemes of the profane?

## Gods, histories, and migrations

Of the 100 groves analysed for this chapter, I was only able to get the history of 21 shrines. Many of these stories refer to the idols found while digging (for yams), or bathing in the nearby stream, or which were brought from across the state – Odisha to Bastar or vice versa, and from Telengana to Bastar (the Telengana deities are said to come from Orangal, a corruption of the term Warangal). Some of these distances are vast and in earlier days would have taken several days, even weeks, of travel on foot.

There are instances of personal relations between the goddesses and the people as with the stealing of the anklet of Kodgudin Mata, whose curse brought about an illness on the people of Baluguda (a village in Bastar – Kodgudin Mata was taken from here to Bansuli in Nabrangpur district, Odisha) until they repented and made peace-offerings, upon which calm was restored. The goddess Birunpalin was refused refuge in Chindawada and Darba, both places not very far from her home, and had to return to Padarpara, a hamlet on the outskirts of her home village, Birunpal.

The priest of Belputi (Bastar) who installed an idol of Kodoi Mata he found while bathing in the Bainsa Badna stream was enraged at her when his children were afflicted with smallpox and died. He smashed the idol and made off to Bajargarh (Nabrangpur), leaving the Belputi people to tend to the shrine without a priest. Fortunately, the pieces of the idol came back together again and a new priest was found. Such instances determine relations between village communities today. Hostility, suspicion, and friendship are some of the legacies that have been passed down.

Other curious instances abound. The goddess of Dewur was brought from Madanar and one day, the king of Bastar, Prabir Chandra Bhanjadev, came by on a visit. He wanted to know who the goddess was in the shrine and when told of her identity, he insisted on offering worship. His offerings were not accepted by her and she responded by giving him a severe itch with which he went off to his palace to Jagdalpur! In the Kangoli shrine, the people found a stream below the ground they were digging. They also found a rice-pounder, a hearth, and vessels in which to cook. Apparently, the gods come to life at night and cook and have their meals.

The presence of the Durwa people in Soutnar and Tehekpal came about as follows – the deity in Soutnar was brought by the Durwa from Orangal with some people on horseback. In the forest near Soutnar, the horse died – this is the spot where the present shrine and grove are – and the man who rode the horse stayed back. He lived in the forest, hunting, and gathering his food, until the people of Soutnar took him in and he settled down there. The goddess of Tehekpal was brought by the Durwa people from Sitapur, who also settled down there.

## The profane

> **profane** [ME. Via Old French from] Latin *profanus*, literally 'outside the temple', from *fanum* 'temple' (whence FANATIC)] Ritually unclean; to defile something sacred or deserving respect (15c); irreverent, blasphemous, hence **profanity** profane behaviour or language (17c, rare before 19c).
>
> (McDonald 2010)

Despite the array of positive influences that a sacred grove has on the community that nurtures them, as well as the people from nearby villages who visit it, almost all the groves have faced some form of violation. Except for eight of the groves (out of a hundred), the rest are in revenue lands and are controlled by communities and not the Forest Department. (Some groves, as the one in Sandh Karmari, is on land known as 'Un-demarcated forest' or 'Orange Area'. These are disputed lands – the dispute for ownership being between the forest and revenue departments.) This means that the community decides and cares for the grove, where the state of the grove is the direct responsibility of the people who live in the vicinity.

Some common reasons for loss of the grove, or its dwindling, are wind and rain, as many groves are located at the edge of streams or rivers that overflow seasonally and erode the grove, causing older trees to fall over. In some villages, cattle are allowed to graze within the grove and in one, the cattle–pen was constructed next to the grove. In another instance, the village pond was located within the grove that brought a large number of people into the space every day. People in another village mentioned that their grove was destroyed by the Forest Department. But in most cases the desecrators are the communities themselves, the local residents, who steal wood, encroach into the grove to extend their rice fields and, in a few cases, succumb to the influence of powerful landlords and allow them to usurp the grove!

Damage to the groves often begins with a road cutting across it, which leads to an excess of human traffic. Human traffic may not necessarily be a bad thing, but people who have not imbibed the deeper spiritual values that brought about the existence of such groves and have not nurtured them, may not see the grove as it is meant to be seen. Outsiders do not easily take in the spiritual legacy of the lore that speaks of three women going to dig yams, finding an idol in their basket, bringing it home and falling ill, reinstating it in a shrine by the name of the idol, *Kanda-khai Mata*, and then having a grove around the shrine today that spreads over a hundred acres. Most people see the physical structure alone, of medicinal plants that can be harvested, of larger trees that can be sold as timber, of the grove as a physical impediment to access the 'other side'. Therefore, they demand and succeed in having a road built through it. This is the situation in both Koleng and Sandh Karmari, the finest of groves in the Bastar district, where a road cuts through the middle, lessening the majestic stature that the sheer size and structure of the groves once commanded.

## Conclusion

Yams, idols, Orangal, illnesses, anger, fear, gratitude, and the supernatural are recurring themes in the non-tangible concerns around sacred groves and their presence among people. For the local communities these are at least equally, if not the far more imposing aspects of sacred groves. When one mentions a grove, it is more likely that a local person thinks of the goddess (Mauli Mata, Kotgudin) who resides there, the kinds of illnesses she cures, and the sacrifices made there periodically, rather than the unusual or rare plant found there. I propose that these aspects – the intangible, the metaphysical – have a bearing on the physical and day-to-day well-being of the groves.

We now revisit the fence of faith referred to earlier. These are fences that have been constructed over generations using the mortar of the intangible and the lime of the very physical aspects of a forest. The modern perception of groves, and the forest space in general, does not include the former component that can conceive of idols of goddesses in their yam basket. Nor is it the intention to create a space that no conservation movement may come within a mile of realizing. Unfortunately, it is our dealings with the 'rational' – in the form of school and education, markets and road access, and a department that claims to manage forests – that has breached the fence protecting sacred groves. It is challenging to disentangle ourselves from a way of thought that we believe in, but we know quite well that barbed-wire fences will not protect our groves.

The state of sacred groves reflects the state of coherence within the Adivasi communities that once nurtured them – this much is clear from the analysis. Today, the people's mundane everyday preoccupations – necessary though they might be – have allowed the trespassing of sacred spaces to continue, profaning them without a second thought. This has been made possible by adapting to the ideas and values of the larger mainstream society for a material advantage, in many instances justified by convenient economic and legal logic. As conservation is understood and viewed through quantifiable parameters alone, such as the *area* of a stand of forest, *biodiversity*, and ecosystem services, the holistic aspects of sacred groves tend to fall out of state and academic purview. The kind of 'conservation' seen in a sacred grove is only *in situ*, and only a small part of it – each instance is unique and will remain so. These groves are neither replicable nor replaceable in quantitative terms. Perhaps understanding and respecting is the best way to protect them.

## References

McDonald, F. (2010). *The Penguin book of word histories*. Penguin UK International Edition.

Watkins, C. (2011). Sak. In C. Watkins (Ed.), *The American Heritage Dictionary of Indo-European roots*. Collins Reference.

# Part V
# Conclusion

# 14 Sustainable use and biodiversity conservation

## Experiences, challenges, and ways forward

*Anita Varghese, Snehlata Nath, Meera Anna Oommen, and Mridula Mary Paul*

### Introduction

When we started to bring together this collection of writings, we wanted to explore the need for developing a broader understanding of sustainable use, keeping in mind the global and national policy frameworks that are prevalent. Sustainable use is often discussed in the context of livelihoods of marginalised people dependent on a diminishing resource. Our effort through this book has been to explore a less discussed aspect of sustainable use – 'under what conditions does sustainable use contribute to biodiversity conservation?'

Consumptive and non-consumptive are the two broad categories that define sustainable use, and within the 'consumptive' category are lethal and non-lethal forms of use. Unsustainable use under any of these categories will impact the long-term availability of the natural resource as well as the lives and livelihoods of people dependent on them. In the absence of proper management, guided by traditional knowledge and modern scientific inputs, any form of use can turn unsustainable in either the short or long term, depending on the resource type. Through the experiences of several case studies as outlined in this book, we come to understand that sustainable use is an important paradigm that helps to establish conditions for use and pre-empt degradation or destruction while including different types of stakeholders.

Biodiversity loss and species extinction are indisputably irreversible. Certain human actions and their impacts at several scales are direct causes for these losses and in many cases accelerate the decline. The broader scales of impact, beyond species or habitats used, are difficult to assess, such as pollinators, dispersers, ecosystem services, and functions (Ticktin 2004). Additionally, the impacts of direct use of a resource may be exacerbated further by erratic climate, natural disasters, zoonotic diseases, etc. The socio-political-cultural context may also change drastically and lead to additional impacts. It seems the simplest way to safeguard then is to ban or stop all forms of use in order to prevent any losses. Is this a real solution? Is it socially acceptable? Do bans really work or do they open avenues for illegal pathways to use and trade? Are there other options?

DOI: 10.4324/9781003343493-19

The chapters we have included in this collection illustrate several narratives of the sustainable use paradigm. Balancing sustainable use and biodiversity conservation is incredibly complex. In the introduction to the book, we created a matrix of various sustainable use categories (Table 1.1). The chapters were planned with the aim to have at least one example for each category. NTFP harvest, medicinal plant collection, fishing, hunting, and grazing are some of the consumptive uses that are documented through the chapters. The latter two are not commonly discussed in sustainable use literature in the context of India. We deliberately brought in these chapters to enrich our outlook and understanding of issues which are normally seen as taboo. Non-consumptive uses such as nature education, eco-tourism, citizen science programs, and faith-based conservation are also important parts of the sustainable use narrative. In this chapter, we assimilate learnings from the previous chapters and case studies.

## Creating a shared understanding

Foremost to creating a shared understanding would be to answer this question – who are the users? Primary users such as indigenous communities, local people, and citizens often interact directly with the resource. This may be through the roles of harvesters, grazers, hunters, fishers, ecosystem service beneficiaries, tourists, or other cultural practitioners. What we see though is that when it comes to creating knowledge, the manager, academic, and policymaker are often the ones who play primary roles here. In many cases, conversations with the primary users do occur, though further exploration is needed to determine whether that is the norm. Therefore, the role of NGOs and academics in facilitating the process of building a shared understanding cannot be overlooked.

A key issue in terms of natural resources that needs to be addressed is the question of power, since the one who controls the resource is rarely the one who uses it. When it comes to wildlife management especially, even researchers and academics have only recently started to take the local people's perspectives into account (Aiyadurai, A., Chapter 8). Most of the early perceptions of resource use saw local people solely as resource users who degrade natural environments. The idea that the users have knowledge and are 'managing' the resource because their livelihoods depend on it was overlooked. Keeping opportunities open to incorporate more disciplines in our research is important, especially in human use landscapes.

Scientific management has not always helped to solve the problem. For instance, fisheries management in Lakshadweep is almost pushed to unsustainable limits when management plans are made with a singular focus on profit, without incorporating local knowledge. Involving local communities in monitoring, management, and decision-making has been critical for the sustainability of traditional fisheries on these islands (Namboothri, N., Chapter 4). Hunting, fishing, and grazing have always been seen as depleting and degrading the resource, however, as portrayed in the example of the Banni Grasslands, unless we create an understanding of prevalent practices and knowledge, we will not have the complete picture (Joshi, P., Chapter 9).

Ethnographies provide rich histories of sustainable use among communities and point to their location within a larger cosmology. In attempting to tease out only the ecologically relevant elements and seeking out implementable strategies, we miss out on the larger value systems and behavioural changes that the belief systems hold (Oommen, M.A., Chapter 7 and Ramnath, M., Chapter 13)

## Making community-based approaches effective

While 'community-based' has become a mantra of sorts, bringing to mind images of women, men, elders, and youths sitting together with managers and arriving at a solution, in the reality of practice this is seldom the case. Even when this does happen, although it may be initially very well-received, time and dwindling incentives make it increasingly difficult to sustain motivation and interest. One also wonders whether it is fair to add these burdens on communities who have very little power in the larger scheme of managing the resource.

Another factor is the composition of hamlets or villages and the challenges posed by heterogeneity with respect to caste, economic class, and gender. Even in the case of homogenous communities, there may be several different leaders and disparate groups of households that owe allegiance to each of them. This is not to discourage community-based initiatives – decades of work in various geographies have shown workable community-based models. It would be unwise to throw the baby out with the bath water.

Communities of practice who have interacted with an ecological system for generations show greater responsibility in keeping the system going since their lives depend on it. In the case of medicinal plant trade which has large turnovers, a carefully planned protocol is necessary to organise the process (Jagannatha, R., Chapter 6). This is not to say that the relationship of communities to the resource is only guided by utility; however, this may be the biggest built-in incentive for conservation. Additionally, as we see in the case of the marine turtle monitoring program, not only fishers but also schools and citizen groups make up the community that looks after the turtles (Shanker, K., Chapter 10). The need for a long-term commitment to make community-based approaches effective is also highlighted (Nath, S., Chapter 5).

In situations such as the Andaman Islands, community profiles have been altered through migrations. Settlers and local administrative bodies need to be cognisant of the value base from which indigenous practices originate and then take proactive steps to incorporate these ways of thinking into conservation efforts (Oommen, M.A., Chapter 7). Community institutions that are proactive can play an important role in keeping good practices alive as we see in the example of the fisheries in Lakshadweep islands (Namboothri,

N., Chapter 4). When a wider constituency of participants who engage with nature (e.g. tourists, holiday makers, school children, concerned citizens) are taken into consideration, there is potential for building support for models of non-consumptive use that go beyond local communities (Whitaker, Z., Chapter 11 and Sharma, K., Chapter 12).

## Measuring conservation outcomes

Does biodiversity benefit from human use? This has been a long-standing question that does not have a simple answer. For instance, many traditional practices in terrestrial ecosystems, such as the setting of low-intensity ground fires in the dry season to increase visibility for hunting, also stimulate re-sprouting of grasses for cattle, keep tick populations in check, control invasive species, and keep paths open for easier movement (Berkes 2017). However, concurrent anthropogenic disturbances confounded with the intensity of disturbance, the biology of the species present, and variations in habitat conditions make it difficult to assess the flow of benefits. In reality, what we have is a set of studies across different landscapes, which are used to benchmark decisions that need to be made for specific landscapes. Unless we have an integrated and interdisciplinary perspective on the complexity of human use and conservation outcomes, evidence for the mutually beneficial relationship between the two is on thin ice.

In an ideal scenario of sustainable use, the rate of harvest or use must be lower than the rate of growth or renewal of the resource. This is often not possible if the species itself is endangered. In such cases, an intervention from the outside is crucial to turn the tide towards a more positive conservation outcome (Aiyadurai, A., Chapter 8). If the status of the species is to be measured, we need baselines and therein lies the challenge, compounded by shifts in flowering patterns, length of seasons, new migratory paths, and unexpected events such as natural disasters. The need for regular monitoring with participatory practices can perhaps contribute to improved baselines (Shanker, K., Chapter 10).

In the case of wild honey collection in the Nilgiri Biosphere Reserve, we see that in the absence of any baseline regarding the status of wild bee populations in the forest, only the community of honey hunters have first-hand information (Nath, S., Chapter 5). Grazing in the Banni Grasslands is an old practice and the Maldharis with their traditional management have kept the grasslands in a state of protection (Joshi, P., Chapter 9). Matching the outcomes of anthropocentric management for use with scientific management for species preservation has to come to a middle ground.

## Good practices of shared responsibility

Long-term partnerships between communities, government agencies, civil society organisations, and other relevant institutions need to be relooked at and strengthened. When issues such as tenure, use, and monitoring sit within the ambit of multiple separate agencies, then the interests for each stakeholder are different which leads to tensions. The issue of power imbalances between stakeholders needs to be addressed if conservation outcomes are to be driven through the sustainable use framework (Paul, M., Chapter 3).

Perhaps these shared responsibilities are also better seen through shared resources, such as NTFPs that are important to communities and also are part of the shared forest ecosystem. A sustainable use strategy for biodiversity conservation with an NTFP as the focal species can be implemented in relevant landscapes (Nath, S., Chapter 5). Charismatic species have an important role to play at national and global levels and perhaps, connections can be made even at the level of states and regions. After all, the honey bee is linked to the tiger in many ways. The establishment of the Madras Crocodile Bank is one such example (Whitaker, Z., Chapter 11). With the turtle monitoring program one also realises that charisma can be linked to other aspects of ecology such as evolution and ancient species. Monitoring programs also give communities the opportunity to engage with science, understand it, and use the language to negotiate (Shanker, K., Chapter 10).

The value of the sacred and cultural aspects of use and how that impacts human behaviour must be acknowledged when bringing various stakeholders on board new projects. As Ramnath, M (Chapter 13) explains, the sacred grove itself exhorts human conservation behaviours, so accounting for these types of influence requires more ethnographic and historical analysis. This is also highlighted in Oommen's (Chapter 7) case study of the hunting practices in the Andamans. The role of local cultural taboos in maintaining certain behaviours and relationships with the resource is often overlooked when management plans and scientific methods are applied.

The case of fisheries is important to add, because when industrialised fishing operations were introduced and the government began to play a larger role, local knowledge was sidelined for many years and has only been revisited within recent decades (Namboothri, N., Chapter 4). Other relationships with the environment that outsiders such as ecotourists share with a place or a landscape can also be considered as a part of cultural values that are linked to the sustainability of rural spaces (Sharma, K., Chapter 12).

## Suggestions and way forward

Through learning from the very diverse chapters in this book, ranging from species to habitats to varying practices such as gathering, hunting, grazing, recreation, and citizen monitoring, we see some directions emerging. We are closer to understanding those conditions which allow for sustainable use and biodiversity conservation, which we set out to discover, and have listed our conclusions here:

- Good governance mechanisms are a key factor in balancing use and conservation, although how much they influence management decisions is not understood and requires further exploration.

- Natural resource-based enterprises have experiences that could lead the way and show promise – since they are often limited to geographies, communities, and local governance systems, replicability and scaling poses challenges.
- There are many forms of sustainable use prevalent within the diverse social and ecological contexts in India and high potential exists to incorporate these into management decisions.
- Knowledge and practice of sustainable use among communities, researchers, managers, and decision makers is plentiful. However, more work is needed to balance power dynamics, especially between traditional users and knowledge holders, and managers and researchers.
- Uses within the realm of education, recreation, and faith have considerable influence on human behaviour and respect towards nature.

There is a need for a more inclusive and interdisciplinary approach in research and practice that can build understanding across scales and levels. The honey bee helps the tiger, which will ensure the health of the ecosystem, which in turn enables honey bee populations to thrive – this cycle can also turn in the reverse direction. Looking for solutions through various lenses and in partnership with several stakeholders is key to a win-win solution. Expanding the definition of 'use' to incorporate different types of benefits that accrue from nature to humanity will expand the notion of 'communities' and in turn expand the stage for conservation benefits. Conservation will then cease to be the priority of only the elite few and spread to those that interact with nature every day.

India is strategically placed to become a leader in linking sustainable use with biodiversity conservation. We have on both sides of the divide equally strong legislation and advocates. The time has come to work together, as nations and people around the world are forced to recognize that biodiversity loss does not just mean that a few frogs or orchids will go missing, but that everyday life can be disrupted and rendered unrecognisable. Perhaps this latest lesson that threatened all of mankind may influence those who make conservation decisions to concede that, for better or for worse, we are truly in this all together.

## References

Berkes, F. (2017). *Sacred ecology*. Routledge.

Ticktin, T. (2004). Dynamics of harvested populations of the tropical understory herb Aechmea magdalenae in old-growth versus secondary forests. Biological Conservation, 120(4), 461–470. DOI:10.1016/j.biocon.2004.03.019

# Glossary

Adivasi
Acres
Crore
Feet/foot
Gram sabha
Indigenous
International Union for the Conservation of Nature (IUCN)
Lakh
Panchayat
Scheduled tribe
Tribal

# Index

Note: Page numbers in *italics* indicates figures and page numbers in **bold** indicates tables on the corresponding page.

For Product Safety Concerns and Information please contact our EU
representative GPSR@taylorandfrancis.com
Taylor & Francis Verlag GmbH, Kaufingerstraße 24, 80331 München, Germany

www.ingramcontent.com/pod-product-compliance
Ingram Content Group UK Ltd.
Pitfield, Milton Keynes, MK11 3LW, UK
UKHW020139200926
475702UK00005B/826

* 9 7 8 1 0 3 2 2 9 0 8 3 6 *